FEB 17 2009

THE U.S. ARMY AND IRREGULAR WARFARE, 1775–2007

Selected Papers from the 2007 Conference of Army Historians

Edited by
Richard G. Davis

D1520663

CENTER OF MILITARY HISTORY
UNITED STATES ARMY
WASHINGTON, D.C., 2008

Library of Congress Cataloging-in-Publication Data

Conference of Army Historians (16th : 2007)
The U.S. Army and irregular warfare, 1775–2007 : selected papers from the 2007 Conference of Army Historians / edited by Richard G. Davis.
p. cm.
1. Special forces (Military science)—United States—History—Congresses. 2. Counterinsurgency—United States—History—Congresses. 3. Guerrilla warfare—United States—History—Congresses. I. Davis, Richard G. II. Title.

UA34.C66 2007
355.02'180973—dc22

2008026508

First Printing

CMH Pub 70–111–1

For sale by the Superintendent of Documents, U.S. Government Printing Office
Internet: bookstore.gpo.gov Phone: toll free (866) 512-1800; DC area (202) 512-1800
Fax: (202) 512-2104 Mail: Stop IDCC, Washington, DC 20402-0001

ISBN 978-0-16-081233-0

Contents

Page

Part Three—U.S. Counterinsurgency Operations

Foreword

Using the theme "The U.S. Army and Irregular Warfare, 1775–2007," the 2007 Conference of Army Historians featured over sixty formal papers exploring the nature of unconventional warfare and its significance throughout history. The event also included several workshops and sessions on administrative issues of common concern across the Army historical community and much informal discussion and networking. From the opening of the conference to the final dinner, where U.S. Army Vice Chief of Staff General Richard A. Cody addressed the current operating environment and answered hard questions from the audience about the future, the conference informed participants about differing aspects of the type of warfare confronting the U.S. Army in the Global War on Terrorism today.

The fifteen papers selected for this publication are not only the best of those presented, but they also examine irregular warfare in a wide and diverse range of circumstances and eras. Together, they demonstrate how extremism was intimately connected to this type of warfare and how Americans have, at different times in their history, found themselves acting as insurgents, counterinsurgents, or both. The titles of the papers themselves reflect how often the U.S. Army has engaged in such irregular operations despite a formal focus on conventional warfare. Using imperial British and Italian examples, several presentations also underline how the ease of conquering lands is often no indication of the level of effort required to pacify them and integrate them into a larger whole.

Each paper provides useful insights to the reader, soldiers and historians alike. I hope that after you have finished with this collection you will continue to broaden your knowledge of military history by turning to the U.S. Army Center of Military History (CMH) Web site (http://www.history.army.mil/) and by examining the Center's many publications and other features highlighting the 233-year history of our Army.

Washington, D.C.	JEFFREY J. CLARKE
1 August 2008	Chief of Military History

The U.S. Army and Irregular Warfare 1775–2007

Introduction

In August 2007, the U.S. Army Center of Military History sponsored and hosted the Sixteenth Conference of Army Historians, a biennial history conference attended by members of the Army Historical Program, academics from leading colleges and universities, and international scholars from allied nations. This conference serves as a forum and professional development venue for Department of the Army civilian and uniformed historians and encourages institutional cohesion within the Army history community. It was a highly successful conference with greater attendance levels (197 attended and 66 papers presented) than ever before. Even more important, the Conference of Army Historians is designed to meet the concerns of Army leaders by addressing the existing needs of the service. The 2007 conference proved no exception to this practice with its highly relevant topic of "The U.S. Army and Irregular Warfare, 1775–2007."

Out of the many excellent papers presented at the conference, the Center selected fifteen of the best for publication. These papers fall into three categories: those detailing non-American counterinsurgency operations; those concerning special aspects of irregular warfare; and those analyzing U.S. counterinsurgency operations. This introduction seeks to weave these presentations into a more complete picture.

The six papers on non-American counterinsurgency operations cover a period of 1,300 years from the life of Muhammad, the Prophet of Islam, through the Dutch defeat by nationalist insurgents in 1948 in Indonesia.

In "The Roar of the Lions: The Asymmetric Campaigns of Muhammad," Russell G. Rodgers of U.S. Army Forces Command analyzes how Muhammad consistently employed tactics and methods outside the then-current conventions of his culture's definition of warfare. Muhammad demonstrated how a small group with unifying ideology can triumph over a larger more sophisticated foe that lacked firm unifying principles. Rodgers notes that some of Muhammad's enemies might not have even realized the danger Muhammad's movement posed until it was too late. The author carefully abstains from generalizing the personal experiences of the Prophet, but one can not help but wonder how his ruthless use of assassination and mass executions may have sanctioned a continuing belief among his more militant adherents in the expendability of human life in pursuit of their cause.

Two papers examined British operations in the southern colonies during the War of the American Revolution. "Southern (Dis)Comfort: British Phase IV Operations in South Carolina and Georgia, May–September 1780," by U.S. Army Signal Branch historian, Maj. (Ret.) Steven J. Rauch, and "When

Freedom Wore a Red Coat: How Cornwallis' 1781 Virginia Campaign Threatened the Revolution in Virginia," by Professor Gregory J. W. Urwin of Temple University, illuminate British efforts to return three former colonies to loyalty to the Crown. Rauch focuses on the South Carolina and Georgia border town of Augusta in a period when British conventional forces seemed invincible. However, as Rauch explains, the British adopted what we would call today an ineffective "Phase IV," or posthostilities, program because they failed to understand how four years (1776–1780) of anti-British local Patriot rule had transformed the populace by "outing" and ousting the most vociferous Loyalists and converting much of the population to "passive acceptance" of Patriot dominance. Lulled by intelligence estimates that failed to identify the extent of divided loyalties in the South and that exaggerated the influence and support of refugee Loyalists, the British touched off a civil war in the backcountry by treating prominent Rebels too harshly and by permitting returning Loyalists to seek revenge rather than justice. The intense civil conflict, in turn, undermined the British ability to guarantee the personal security of supporters, which fatally compromised the Crown's capacity to restore its rule in Georgia and South Carolina. British inability to match Phase IV planning and objectives with the reality on the ground compromised the political and strategic end state of the campaign.

Moving north, Urwin examines Lord Cornwallis' operations in Virginia in 1781 with an unusual emphasis on the British relations with the servile African American population of the colony. The Crown offered blacks freedom for loyalty—a simple, compelling, and truly revolutionary concept in a colony with a 40 percent slave population. Virginia's slaves flocked to their liberators, providing the British willing workers, unmatched local intelligence, and a resident population with a relatively high number of completely loyal citizens. Cornwallis took full advantage of this situation. He made Virginia howl, especially its pro-Patriot, slave-owning upper class. Throughout the spring of 1781, he conducted effective conventional operations against inferior American forces coupled with equally effective antipartisan, and antilocal government operations—all of which disrupted Patriot resistance efforts. Had he continued he might have eviscerated Virginia's ability to contribute to the Patriot war effort or even brought it back to the Crown. However, in midsummer, Cornwallis received orders from the British commander-in-chief in America, Sir Henry Clinton, to retire to the coast and cease operations into the interior. Cornwallis began to dig entrenchments at Yorktown on 2 August.

Two takeaways from Urwin's paper are that, first, there is a tipping point in the level of public support necessary to conduct successful counterinsurgency operations. The British lacked sufficient support in the Deep South but may, with the support of the slave population, have achieved it had they more diligently pursued offensive operations in Virginia. And, second, consistent follow-through is required for sustained effective counterinsurgency operations.

Clinton's decision to impose a static posture on Cornwallis' forces lost not only the campaign but the entire war.

Chris Madsen, an associate professor, Department of Defence Studies, Canadian Forces College, delineates the involvement of the Canadian Army in the Anglo-Boer War of 1899–1901 in South Africa. His paper, "Learning the Good and the Bad: Canadian Exposure to British Small War Doctrine in South Africa, 1900–1901," offers some useful insights about the counterinsurgent operations of a small, all-volunteer component of a combined British imperial force. The Canadians, under overall British direction, fought well, but normal attrition and wastage, coupled with a shortage of replacements, constantly reduced their capabilities. By 1901, the Canadian government withdrew its forces.

Madsen noted two items valuable to the study of counterinsurgency. At the end of conventional operations, the senior British leaders in South Africa completely failed to realize that they even faced an on-going insurgent threat. Thus, their initial reactions to continued Boer raids and actions were slow and ad hoc. This inadequate response allowed the insurgents and their supporters (the great bulk of the Boer civilians) to organize for further action and to gain confidence. Next, the British reacted to the insurgents by imposing increasingly brutal and repressive pressure on the Boer civilians, until they had forced most of them into defended enclaves the British referred to as concentration camps, where 15 percent of the total Boer population died. Thanks to overwhelming numbers (450,000 soldiers versus 185,000 Boer men, women, and children) not likely to be repeated in any current counterinsurgency, the British succeeded in ending the insurgency. British counterinsurgency tactics in the Boer War represented an extreme case of separating the insurgents from the support they needed to survive.

Richard Carrier, a senior lecturer in international history and the history of war, Royal Military College of Canada, and Thijs W. Brocades Zaalberg of the Netherlands Institute for Military History supply two descriptions of failed colonial counterinsurgent campaigns. Carrier's paper, "Blindness and Contingencies: Italian Failure in Ethiopia (1936–1940)," analyzes the Italians conquest of Ethiopia and their postwar actions. Carrier notes that after the Italians had occupied the capital and approximately one-quarter of the country their senior leadership, including Benito Mussolini, assumed that the conflict was over. Mussolini and his soldiers had conducted no prior Phase IV planning or, indeed, given any thought to postwar counterinsurgency operations or what policies to apply to the 28 million additional inhabitants of their new empire. Mussolini decreed that the Italians would rule the country directly, without sharing any power with natives. This decision alienated the local elites, not all of whom had supported the previous regime. It ensured that important local governmental and administrative decisions would be made by officials with a limited knowledge of conditions and culture. The attempted assassination of the Italian viceroy in February 1937 prompted the occupiers to begin to purge the surviving Ethiopian elites. All the while, the Italians reduced their overall force levels and substituted native for European manpower. They also imposed a policy of strict racial segregation between the conquerors and the conquered. In late 1937, they appeared to change tactics by selecting a new viceroy, the Duke of Aosta, and instructing him to pacify the country with more humane methods. However, no sooner had the duke issued orders restraining his forces than the newly appointed commander of the Italian armed forces in Ethiopia, General Hugo Cavallero, launched a series of campaigns into unoccupied territory. Not until he left, in May 1939, did the duke have a free hand to use the carrot as well as the stick. He made little progress before World War II reversed the conquest.

The Italian experience in Ethiopia showed the importance of producing thoughtful and accurate Phase IV planning. It may also demonstrate that, unless ethnic cleansing is the goal, overly violent and brutal counterinsurgent tactics are self-defeating, especially when the occupying force does not have the strength to control the country. Also, if the occupiers lack sufficient force, adoption of policies that refuse to share any power with native elites and treats all natives as racial inferiors makes little sense. If one cannot either remove or overawe the native populace with force, then one has no choice but to adopt a policy with some accommodation.

In "The Roots of Dutch Counterinsurgency: Balancing and Integrating Military and Civilian Efforts from Aceh to Uruzgan," Zaalberg looks at current Dutch counterinsurgency tactics, based on nonkinetic options, such as winning the hearts and minds of the populace, rather than killing insurgents, and asks if the current Dutch method continued practices learned in the past. He then analyzes two Dutch colonial campaigns in the Netherlands East Indies, now known as Indonesia: the pacification of Aceh Province, 1873–1910, a success, and the Dutch attempt to reestablish their colony after the end of World War II, 1945–1949, a failure. In Aceh, the Dutch for decades employed heavy road-bound retaliatory columns to sweep the province. These columns targeted the populace. At the turn of the century, a new commander pacified the province. He worked closely with cultural experts, resorted to more flexible tactics to include smaller columns, improved native-manned special forces–type units, and separated the population from the insurgents. In the post–World War II campaign, the Dutch tried to fight a conventional war with a force that was beyond their means and, yet, still too small for the task at hand. Zaalberg concludes that the Dutch Army's current counterinsurgency methods do not follow past practice. As Zaalberg demonstrated, armed force can selectively modify tradition in response to new problems.

Four papers deal with unusual aspects of warfare. In "The History of Military Commissions in the U.S. Army: From the Mexican-American War to the War on Terrorism," Frederic L. Borch, a historian with The Army Judge Advocate General's Legal Center and School, examines the Army's 160-year experience with military commissions, a specialized type of Army judicial proceeding. The Army established its first military commissions in 1847 at the order of Winfield Scott, commanding general of the Army. General Scott created these courts to close a loophole in the then-operative Army Articles of War of 1806, which did not provide for the punishment of civil crimes perpetrated by U.S. soldiers. This was no problem on U.S. soil, where an offender could be handed over to a functioning U.S. civil court, but, in enemy or occupied territory, the Army itself, as General Scott recognized, needed a way to try a soldier or contractor for civil crimes. By the end of the Mexican War, the commissions extended their authority to try Mexicans for offenses against the rules of war. The commissions used military laws of evidence, which were and are far more permissive than those of U.S. civil courts.

After a detailed history of military commissions from the American Civil War through World War II, Borch addressed President George W. Bush's revival of military commissions to judge terrorists. The president strictly limited the commission's jurisdiction to non-U.S. citizens and only for terrorist-related offenses that "threaten to cause, or have as their aim to cause, injury to . . . the United States." In 2006, Congress passed the Military Commissions Act, which defined the commissions and their procedures and allowed evidence produced by hearsay and torture. Borch concludes by noting that the future of military commissions "is very much in doubt." This paper is necessary reading for all citizens in that it presents essential facts concerning these unique judicial institutions. Whether one approves or disapproves of military commissions, this paper gives the reader a basis of knowledge to develop an informed argument.

A second specialized paper, "Intimidation, Provocation, Conspiracy, and Intrigue: The Militias of Kentucky, 1859–1861," from the command historian of the 81st Regional Readiness Command, Lt. Col. John A. Boyd, analyzes the role of the Kentucky state militia in the beginning months of the American Civil War. In 1860, the state militia, designated the Kentucky State Guard (KSG), constituted the armed force available to the state government during the early period of the Civil War in which Kentucky had to decide to remain in the Union or side with the breakaway Confederate states. The mere presence of the state guard posed a threat to the pro-Unionists, who controlled the Kentucky legislature, However, the position of the militia became untenable when President Lincoln authorized the clandestine shipment of small arms to private Union militias, which broke the KSG's military dominance in the state. The new equilibrium emboldened the legislature, which cut off funds to the Kentucky State Guard and required all of its members to take an oath of loyalty to the Union. With the breakup of the state guard, Kentucky remained in the Union. Boyd concludes that the main contributions of the Kentucky State Guard were that it kept the state neutral for an extended period, that it inadvertently admitted anti-Union hotheads into its ranks rather than having them form a more active party of their own, and that the peace that it enforced allowed the partisans of each party to divide peaceably. He might have added that the restraint of the KSG's leaders and their refusal to resort to brutality and strong-arm tactics allowed for a peaceful settlement of the question of Kentucky's relationship with the Union. Most modern states faced with civil war were not so blessed.

In "The Spoliation of Defenseless Farmers and Villagers: The Long-Term Effects of John Hunt Morgan's Raid on an Indiana Community," Stephen I. Rockenbach, an assistant professor of history at Virginia State University, looks at the aftereffects of Morgan's raid on the local perceptions of the Indiana and Kentucky populace on each side of the Ohio River, especially Meade County, Kentucky, and Harrison County, Indiana. In July 1863, Confederate Brig. Gen. John Hunt Morgan, a former company commander in the Kentucky State Guard, crossed the Ohio River into Indiana, and, though his raid caused some immediate destruction and disruption, he was soon trapped and forced to surrender his force. Although the raid had little military value, Rockenbach contends that it had lasting political effects on both sides of the river. In Indiana, the citizenry lost faith in the effectiveness of their military and their state and national governments. They also almost ceased their active support for Unionists south of the river. In Kentucky, Morgan's raid emboldened Confederate sympathizers, who emerged to take local political control and elect pro-southern officials to all levels of government before and after the end of the Civil War. Rockenbach's case study demonstrates how even a single raid or instance of violence has long-lasting consequences. It suggests that further research is needed on the effect of repeated violence and brutality on a population. Violence, like many other actions, has a point of diminishing returns. Consequently, this paper implies that insurgents may prefer to spread violent actions widely but thinly over a region than to concentrate such activities in a few locales. And, if, as seems intuitive, that widely spread insurgent violence is, up to a point, more efficient and possibly more effective than concentrated insurgent violence, what options should those directing the counterinsurgency employ to defeat this activity, especially if they do not possess an overwhelming advantage of forces?

In the fourth specialized paper, "Irregular Engineers: The Use of Indigenous Labor in the Rebuilding of Critical Infrastructure During the Korean War, 1950–1953," Eric A. Sibul, an assistant professor of defense studies at Baltic Defense College, Tartu, Estonia, focuses on the use of local Korean labor in keeping the Korean railroad system operating during the war. Railroads were and are the most efficient method of shipping freight for both military and civilian uses. In Korea, the rail system was the backbone, or key infrastructure, sustaining both United Nations military operations and the civilian economy. During the war, the U.S. Eighth Army, working closely with the Korean National Railroad, the Korean state agency that owned and operated the rail system, employed thousands of Koreans as laborers and many others as engineers and supervisors. As Sibul explains, this scheme benefited both the Koreans and the Eighth Army. It may not have made the locals supporters of their government or its allies, but it kept them productive and out of trouble. Reconstruction of the railroad also gave the population a vested interest in preserving their work, if not their state. The cash the workers earned and the material purchased from the local suppliers turned over several times throughout the Koreans' war damaged economy, indirectly sustaining thousands more. For the Americans, the cheap, productive, and highly cooperative laborers and the surprisingly efficient engineering, enabled them to maintain their operations at a minimum cost in dollars and personnel. Both sides gained. Sibul's paper provides an excellent case study of an aspect of counterinsurgency that is somewhat neglected—the necessity of keeping the civilian population gainfully employed and the local economy liquid and sustained. A man with a shovel in his hand repairing a road is not a man sitting at home with no income and tempted to pick up an AK47 and fire a few shots at the occupiers or their supporters in return for insurrectionist dollars. It would follow from this, although Sibul does not explicitly state it, that the occupier should also ensure that the employment of native labor is seen by the populace as voluntary and fairly compensated. Forced or slave labor would be entirely counterproductive. Sibul also noted a particular American flaw in this aspect of counterinsurgency. American engineers were so wedded to machine-driven, material solutions that they failed to appreciate the possibilities of mass hand-labor to accomplish the same tasks. Phase IV planners would do well to heed the insights offered by Sibul and emphasize large-scale employment of native labor and associated professionals as soon as possible during an occupation period. The results might even generate an unexpected fount of goodwill for the occupier-counterinsurgent.

The five remaining papers examine U.S. Army counterinsurgency operations. In "The Victorio Campaign: Hunting Down an Elusive Enemy," Kendall D. Gott of the U.S. Army Combat Studies Institute analyzes a U.S. Army counterinsurgency operation against a band of Apaches in 1879–1880. As Gott notes, Victorio and his followers had no of chance of ultimate victory. Only the unmatched hardiness of his warriors, their light logistics requirements, and their intimate knowledge of the land allowed them to survive as long as they did. Consequently, the measure of the Army's success in this campaign was not its final expulsion of Victorio into Mexico, but the length of time it took and the expenditure of resources required to achieve its end. Gott finds three key elements to the Army's eventual triumph. First and foremost was the Army's use of Native Americans as scouts. These scouts provided highly accurate intelligence concerning Victorio's location and his intentions. The scouts also tended to negate the Indians' advantage of knowing the terrain. Second, the U.S. commander, Col. Benjamin Grierson, adopted flexible tactics to fit the situation. Instead of launching ineffective offensive sweeps through the countryside, he concentrated on Victorio's logistical weakness, water. The colonel placed guard forces at all water sources, denying them to this enemy. Finally, Grierson kept up a determined and patient pursuit of Victorio. The pursuit, while costly in horses and supplies and time consuming in the face of an exasperated public and press, ultimately wore Victorio's band down to the point where they retreated into Mexico. From this Gott draws three lessons applicable to current operations: "commanders must understand the enemy's methods of operation and exploit his weaknesses," and "commanders and staffs must also look beyond their formal training in devising flexible tactics and strategy, and in preparing their units for sustained operations that [could] last for months."

Given Gott's analysis, one might also observe that counterinsurgent forces should actively recruit local nationals, if possible from tribes or factions opposed to the insurgents, to aid in intelligence gathering, supply cultural information on enemy tactics and intentions, and serve as scouts or guides. In the past, the U.S. Army put Indian and Philippine scouts to good use. Perhaps "Iraqi scouts" assigned to regular units, not just to Special Forces, could prove useful in the war with Iraq. Repugnant as it may be to soldiers to use "collaborationists" and as anxious as the United States is to build a self-sustaining Iraqi Army, it may be a sensible tactic to have some small units of Iraqis subsumed into the U.S. Army itself.

The remaining four papers cover operations in the first thirty years of the Cold War, from the Greek Civil War through the war in Vietnam. Robert M. Mages of the Military History Institute, U.S. Army Heritage and Education Center, looks at the beginning of the period in his paper, "Without the Need of a Single American Rifleman: James Van Fleet and His Lessons Learned as Commander of the Joint United States Military Advisory and Planning Group During the Greek Civil War, 1948–1949." In the late 1940s, the civil war between the Greek royal government and the Greek Communist Party threatened to complete Soviet control of the Balkan Peninsula, to put Turkey into a two-front situation, and to give the Soviets open access to the Mediterranean Sea. When the British announced that they could no longer afford to support the Greek royal government in 1947, the Americans took their place. In early 1948, U.S. Army Lt. Gen. James Van Fleet became head of the Joint U.S. Military Advisory and Planning Group, which directed U.S. assistance to the Greek Army. After disappointing campaigns in 1948 revealed grave operational shortcomings, Van Fleet used the winter of 1948–1949 to more completely train the Greek Army in the American image. In 1949, the Greek Army smashed the insurgents and ended their threat to the government. Van Fleet attributed this victory to four basic principles: work with and support elements in the indigenous armed forces and government that share American goals and objectives; demand accountability; build the indigenous army according to the requirements of the conflict; and remember that the war must be waged and won by the indigenous army and the government it defends.

Van Fleet had done well in reasonably favorable circumstances. He built on the foundation of an already established state and an existing Army, both of which had a measure of legitimacy and support from the Greek people.

In "Chasing a Chameleon: The U.S. Army Counterinsurgency Experience in Korea, 1945–1952," Lt. Col. (Ret.) Mark J. Reardon of the U.S. Army Center of Military History dissects a much less favorable scenario that faced the U.S. Army in South Korea. During the period covered by this study, the political situation and the nature of the insurgency underwent three major changes. From September 1945 through May 1948, the United States occupied South Korea and ruled it through a military government. At this time, the U.S. Army organized and recruited the Korean National Police and Korean Constabulary (which became the Korean Army) to maintain order and suppress opponents. The second period, 1948–1950, began with the founding, on 15 August 1948, of the Republic of Korea (ROK). After some initial difficulties, the republic seemed to be gaining the upper hand over the local southern insurgents. This prompted the North Korean regime to send increased aid and manpower into the South in hopes of overthrowing the republic. During this period, the U.S. Army greatly reduced the number of advisers to the National Police while raising the total number of advisers to the ROK Army from 100 to 248. By 1949, using brutal and violent tactics, the National Police and the ROK Army suppressed the northern-backed insurgents. This, in turn, convinced the northern leadership that only a full-scale military invasion could overthrow the republic. That invasion, in June 1950, touched off the Korean War and led to the third period studied by Reardon, the wartime insurgency of 1950–1952. In early and mid-1951, while Van Fleet, now the commander of Eighth Army, rebuilt, expanded, and trained the Korean Army, American units conducted division-size counterinsurgency operations. Beginning in late 1951, the National Police and ROK Army units, with U.S. advisers and U.S. air and logistical support, conducted successful division- and even corps-size counterinsurgency operations behind the United Nations front lines. Reardon concludes that the example offered by Korea serves to alert U.S. commanders that they must be prepared to assume a counterinsurgency role in the absence of combat-effective indigenous military forces. His paper further demonstrates that the nature of both the insurgency and the counterinsurgency will evolve, sometimes radically, over time. Consequently, tactics and assessments must be flexible and responsive—a lesson the U.S. Army learned well in operations in Afghanistan and Iraq in 2001–2007.

In "Lessons Learned and Relearned: Gun Trucks on Route 19 in Vietnam," Ted Ballard, a retired historian from the U.S. Army Center of Military History, narrowly focuses on the operations of the 54th Transportation Battalion along the 100-mile-long Route 19 in the Central Highlands of Vietnam, from the port of Qui Nhon to the city of Pleiku, with a mission of supplying transportation to the II Corps Tactical Zone. He further concentrates on events in mid- and late 1967 that led members of affected transportation units to develop the "gun truck," a heavily armed and armored five-ton truck designed to act as a convoy defender to drive off attacking insurgents. The U.S. truck convoys needed their own self-defense capabilities along isolated patches of the highway that could not receive immediate support from American tactical units or when those units were unavailable. Ballard demonstrates how tactics on both sides adjusted to each others' responses. The gun truck proved effective for the remainder of the war. And, as the title of his paper suggests. Ballard went on to note that the Army never institutionalized this experience and soon lost the memory of the gun truck's utility in unescorted convoy operations in hostile territory. In Iraq, the Army repeated this experience and came up with the same expedient, aided only at the end of the process by the "rediscovery" of a Vietnam-era gun truck, nicknamed the "Eve of Destruction," at the Army Transportation Museum at Fort Lee, Virginia.

Although Ballard presents an informative paper about ad hoc innovation in the service, one cannot help but wonder about the aspects of counterinsurgency he did not touch on, undoubtedly because of his decision to focus so closely on his subject unit. He never identifies the enemy; were they Viet Cong local insurgents or North Vietnamese regulars, and what were their unit designations? Likewise, Ballard made no mention of the Vietnamese population on either side of the road. Did the insurgents swim in this population? We have no clue. Perhaps, Ballard has fallen victim to the typical American fascination with equipment. Not once does he speak of any attempt of the transportation units to interact with the roadside population. If this is so, it serves as an object lesson in how not to conduct insurgency operations. Ignoring the locals, except to level their fields to expand the open areas on both sides of the road, is not a viable option.

In the final paper of this volume, "Some Observations on Americans Advising Indigenous Forces," Robert D. Ramsey III of the U.S. Army Combat Studies Institute summarizes the U.S. Army's advisory efforts in the Korean War, 1950–1953, the war in Vietnam, 1961–1973, and the civil war in El Salvador in the 1980s. In Korea, the number of American advisers increased more than elevenfold from approximately 250 in the beginning to 2,861 in the end. This mirrored the increase in the size of the ROK Army itself. In Vietnam, at the height of the advisory effort, over 14,000 U.S. Army advisers (seven division equivalents of officers and noncommissioned officers) served with the Army of the Republic of Vietnam and with pacification efforts. In El Salvador, the U.S. Congress limited the advisory effort officially to 55, and they advised a force of only a few thousand soldiers in a handful of brigades as opposed to some twenty divisions in the other conflicts. Ramsey concentrated on the (often depressing) similarities between the three efforts. The average U.S. adviser did not have even a limited understanding of the local language (apparently, the U.S. Army even had trouble fielding Spanish-speaking officers in El Salvador) or culture. This led to an overreliance on translators for communication. Lack of cultural awareness also meant that advisers, on too many occasions, ignored indigenous practice and experience and, instead, advocated American methods and solutions. Only in Vietnam did the Army give its advisers some measure of advance training: a six-week Military Assistance Training Advisor course.

In the other conflicts, if a soldier was branch qualified (had a basic knowledge of his Army specialty), he was generally accepted as an adviser and provided little or no additional training. By design, an adviser was intended to be one rank below his local-nation counterpart, but as the conflict continued the American adviser was often two or more ranks junior. Unlike their host-nation equals, the advisers served relatively short tours. In practice all this meant that the local indigenous commander, who had a good deal more on his plate than just dealing with his advisers, was confronted with a stream of cock-sure young Americans who knew little of his culture or language and who wanted to be his "friend" and to tell him how to fight like an American. In conclusion, Ramsey observed,

> First, it is a difficult job. Working effectively with indigenous forces may be the most difficult military task. Second, not everyone can do advisory duty well. Advisers should be carefully screened and selected. Third, advisers should receive in-depth training about the host nation, its history, its culture, and its language. Without such training, situational awareness is almost impossible. Fourth, adviser training should focus on advisory duties and harnessing host-nation institutions, organizations, procedures, capabilities, and limitations. It should not focus on U.S. solutions to host-nation problems. Fifth, longer, repetitive tours by specially selected and well-trained advisers enhance the development of rapport with host-nation counterparts. Lastly, no matter how capable field advisers are, their success depends upon the support structure established between the host nation and their U.S. chain of command.

These are useful insights, but ones likely to be ignored by the bureaucratic processes of the modern U.S. Army. In particular, the personnel system discourages long-term or repeat assignments to the same duties, while the Army itself cannot afford to parcel out specialized intense training to large numbers of soldiers on the chance that they might have to serve as advisers in unforeseen countries at some undetermined future dates. In 2000, could the Army have anticipated needing (and pooling in advance) several hundred advisers for Afghanistan (not to mention thousands of advisers for Iraq)?

Finally, one must ask what is the measurement of success for an advisory effort? Is it, at a minimum, the creation of a competent local fighting force that can effectively defend its state from internal insurgencies and, with assistance from allies (the United States), repel any foreign invader? In Greece and Korea, U.S. advisers met that definition, although they also left behind armies that eventually suppressed the governments they were supposed to defend and ruled autocratically in their stead. Both these forces also continued to receive U.S. weapons and military aid for a considerable period after the end of hostilities. In Vietnam, the advisers helped to produce a force that could fight with U.S. assistance and support, but not without it. In El Salvador, the U.S. advisers helped to field a force that successfully defended the established government and faced no external threat but created a force that was also guilty, along the way, of some serious civil rights abuses.

The selected papers of the 2007 Conference of Army Historians reflect the diversity of the contributors and the complexity of the subject—irregular warfare. No two of them are directly comparable and no two of them address the same subject. Yet, taken as a whole, they offer the reader insight into the unique circumstances of each counterinsurgency examined. If there is one lesson to take from this collection, it is that the leadership of a counterinsurgency operation or program must have an accurate understanding of both the weaknesses and strengths of the enemy and of the expectations and needs of the populace in order to stand a chance of success.

PART ONE

Non-American Counterinsurgency Operations

The Roar of Lions: The Asymmetric Campaigns of Muhammad

Russell G. Rodgers

You loathed Medina when you visited it, and met roaring lions there.

—Hassan bin Thabit, Muhammad's poet.
Quoted in Ibn Ishaq, *The Life of Muhammad*, p. 489.

One of the problems when examining asymmetric warfare is defining it. While definitions abound, few cut to the chase of what such warfare truly is. The key problem confronting the definitions is that it often confuses technological change with asymmetric warfare, failing to note that such warfare has existed since the dawn of history. While technology is ever changing, the techniques of asymmetric warfare are ancient. With that in mind, what is required to fight in such an environment is not a new paradigm, but a paradigm digression; that is, an application of what worked from the past. To understand this, one must first know the past, especially of the enemy, and how he fought an asymmetric campaign. Even more telling is how his opponents failed to defeat him.

The Prophet Muhammad's campaigns in western Arabia require some background information to understand his objectives and why he chose the tactics he did to achieve them. Arabia in the sixth and seventh centuries developed largely in the shadows of two massive empires: the Byzantine to the northwest and the Persian, known at that time as the Sassanid, to the northeast. Byzantium was essentially a Christian empire while the Sassanid was Zoroastrian, or what early Islam and Christianity would call pagan. These two empires struggled for control of the Fertile Crescent, the land from Palestine to Mesopotamia, in an effort to dominate trade with the east. Byzantine merchants plying these trade routes grew weary of the high tariffs imposed by the Sassanid governors and sought alternate routes, which they found in the desert of Arabia.[1] Such trade routes transformed Makkah, which had once been a minor village with a watering site, into a thriving merchant and religious community that dominated the region.

Therefore, control of Makkah became increasingly important from the third through the fifth centuries. Several tribes jockeyed to control its growing

---

[1] W. Montgomery Watt, *Muhammad at Mecca* (Oxford, UK: Oxford University Press, 2004), pp. 11–12.

wealth and importance, with the tribe of the Quraysh finally gaining the upper hand. Qusayy, called the King of the Quraysh, pulled the reins of power into the hands of his small elite around 400 AD. Having done so, the Quraysh so dominated economic and social life that early Islamic historians called his rule a "religious law."[2] In this context it is important to understand that in ancient Arabic the word for *religion* is the same for *obedience*.[3] Thus, religion for the Arabic world was much different than what many in the West today would think of it. In their world it meant total obedience and submission in all areas of life, and, under the early Quraysh, this meant obedience to Qusayy and his close associates. Along with this came the increasingly important focus of religious ideals centered on the Ka'bah, an ancient cubed structure that was the locus of worship for not only Makkans, but tribes from the surrounding area as well.

As Makkah grew and prospered, significant changes began to impact the rule of the Quraysh. First, there was a growing level of division within the tribe itself. Over a 100-year period, a ruling elite developed within the Quraysh centered on specific descendents of Qusayy, specifically the Abdul Dar clan. When Qusayy died, he delegated the five principal offices of Makkah to his eldest son, and these offices controlled the political, economic, and social life of the town. Of these five offices, the most important was *rifahdah*. This involved providing food to religious pilgrims who arrived at Makkah for the annual *hajj*, an event gaining increasing importance parallel to Makkah's rise in mercantile wealth. This, in turn, required the imposition of a property tax on the rest of the Quraysh to defray the expenses. While there were other taxes in Makkah, the *rifahdah* was the most important, being the lion's share of tax revenue.[4] Whoever controlled this, manipulated welfare largess and, ultimately, the political life of Makkah and the surrounding area. Control of these resources led to the development of a princely class that could control the town *mala*, or council, and determine law and policy.

The second involved the rising tide of what we would today call democracy. This developed because of the rising level of wealth among specific merchants, and not just merchants of the inner ruling elite. Within early Arabic societies, the tribe was the all-important social and legal body. Tribal chieftains, or *shaykhs*, steered the life of their subtribes, or clans. They offered direction as well as legal guidance and, more importantly, protection,

---

[2] Muhammad Ibn Ishaq, *The Life of Muhammad,* trans. A. Guillaume (Oxford, UK: Oxford University Press, 2004), p. 53; Abu Ja'far Muhammad bin Jarir al-Tabari, *The History of al-Tabari*, vol. 6, *Muhammad at Mecca*, trans. W. Montgomery Watt and M. V. McDonald (Albany: State University of New York Press, 1988), p. 1098. After making specific references to separate volumes of al-Tabari's history, page numbers organized to the original Arabic and included in the translations will be used, rather than references to separate volumes.

[3] Abdul Mannan 'Omar, *Dictionary of the Holy Qur'an* (Hochessin, Del.: NOOR Foundation-International, 2005), p. 185. The word, pronounced *deen*, is hard to translate as it has numerous shades of meaning, underscoring the difficulties of classical Arabic.

[4] Ibn Ishaq, *The Life of Muhammad*, p. 55.

thus bringing stability to the region. As members of the Quraysh engaged in trading enterprises between Byzantium and Sassanid, a new class began to emerge in the area around Makkah—the moneyed middle class. A propertied middle class was concurrently developing in other parts of Arabia where some tribes began to settle down to a profit-based agriculture. But a moneyed middle class was different. Those that grew wealthy in liquid assets began to assert a significant degree of autonomy over their tribal *shaykhs*, thus threatening the latter's authority.[5]

Connected to this new middle class was a rising tide of religious democracy. Tribes had their gods, represented in idols, but now each clan and family began to select their own gods on which to base their clan's history, stability, and power. These idols, 360 in all, matched the days of the lunar calendar with the addition of a few extras that were considered higher deities.[6] While it would be tempting to compare this to a western pantheon, such would be in error as most of these gods were considered equals. Only five stood out above the rest: al-Uzza, al-Lat, Manat, Hubal, and Allah. While Hubal was considered by some to be the highest of the Ka'bah, it was by Allah that people prior to Muhammad's day made their oaths while standing next to the image of Hubal, thereby indicating Allah's superior status.[7] Because the various other gods were equals, each clan within the Quraysh, as well as the outlying tribes in Arabia, could attempt to claim equal status. Thus these two issues, the rising moneyed middle class and the tide of religious democracy, were serious threats to the power of the elite within the Quraysh.

Indeed, around 450 AD, there was a significant challenge to the ruling element of the Abdul Dar clan by brothers within the tribe. This came from the second son of Qusayy, Abid Manef, now demanding the right to hold the offices of the Ka'bah. A civil war broke out and nearly came to serious fighting within the holy precinct around Makkah, when at last a truce was called and an agreement made. While Abdul Dar retained the lesser of the offices, the Abid Manef, particularly the clan of Hashim, gained the *rifahdah*, shifting the balance of power in Makkah. It was from the princely clan of Hashim, part of the Abid Manef, that Muhammad emerged.

Born in 570 AD, Muhammad was orphaned by the time he was six years old, being raised first by his paternal grandfather and later his uncle Abu Talib. While Abu Talib and the Hashim clan provided Muhammad with protection and position, he was still an orphan, and thus his inheritance—and claim to any leadership—was seriously threatened. Muhammad had to sense this keenly

[5] *Muhammad at Mecca*, p. 19.

[6] The lunar calendar has approximately 354 days.

[7] Ibn Ishaq, *The Life of Muhammad*, pp. 64–67; *Al-Tabari*, p. 1075. Some have noted that Muhammad never made any references to Hubal in the Qur'an, thus generating speculation by some that Hubal and Allah are one and the same. There is little evidence to support such a claim. See Thomas Patrick Hughes, *Dictionary of Islam* (repr., Chicago, Ill.: Kazi Publications, 1994), p. 181.

when he began to engage in merchant trade under his uncle. By the time he was twenty-five, he was still unmarried, having very little in assets to contract for a wife. He at last struck a deal with an older wealthy noblewoman named Khadijah and the two were married in 595. She was fifteen years his senior.

With his marriage into a good family and access to his wife's wealth, Muhammad was able to develop an economic and political base for himself. He began to assert himself within the community, such assertion coming to a head when he proclaimed his prophetic calling for the god Allah when he was forty years old. His first converts were within his own family, a parallel similar to other revolutionaries such as Mao Tse-tung.[8] After that, he received the support of a rising merchant named Abu Bakr, who not only committed financial resources, but recruited others from outside the elite. In fact, Abu Bakr's support was crucial to Muhammad's movement, raising the question as to whether or not he may have been the true power behind the movement.[9]

At this point, it must be understood what Muhammad was asking of his peers and elders within the Quraysh when he called them to Islam. While typically explained as peace and submission to Allah, in fact Islam is more consonant with temporal political submission, which was revealed by the way it was applied.[10] Thus, when Muhammad called those around him to Islam, he was calling on them to give submission primarily to himself and not to Allah. One episode that clearly explains this occurred in Makkah prior to the Prophet's migration to Madinah. Muhammad called a number of the leaders of the Quraysh to have a meal with him, after which he called them to Islam. Nobody stirred until finally young, scrawny Ali, son of Abu Talib, rose up to say he would support the Prophet. Muhammad placed his hand on the back of Ali's neck and said that the young man would be his brother and successor. The latter in Arabic is *khalifa*, and thus was laid the seeds of the later Sunni-Shi'ia controversy. When the elders heard this, they turned to Abu Talib in astonishment and said, "He has commanded you to listen to your son and to obey him!"[11] These elders understood exactly what Islam meant, being the submission tendered first and foremost to the man Muhammad.[12]

---

[8] Jung Chang and Jon Halliday, *Mao: The Untold Story* (New York: Alfred A. Knopf, 2005), pp. 28–29.

[9] There are hints as to the level of his authority and influence, not to mention that he was the first *khalifa* after Muhammad's death. See *The Meanings of Sahih al-Bukhari*, trans. Muhammad Muhsin Khan (New Delhi, 1987), vol. 4, no. 94. Also see an item of poetry by a Qurayshi after a Muslim raid in Ibn Ishaq, *The Life of Muhammad*, p. 283.

[10] Afzalur Rahman, *Muhammad as a Military Leader* (London: Muslim Schools Trust, 1980), pp. 11–12; *Dictionary of the Holy Qur'an*, p. 267.

[11] *Al-Tabari*, pp. 1172–73.

[12] There are numerous examples to illustrate this point. See *Al-Bukhari*, vol. 5, no. 425; Shaykh Safiur-Rahman al-Mubarakpuri, ed., *Tafsir Ibn Kathir* (Riyadh, Saudi Arabia: Darussalam, 2003), 4:396; "Letter to the People of Tihama Range," in *Letters of the Holy Prophet Muhammad*, ed. Sultan Ahmed Qureshi (repr., New Delhi: Kitabbhavan, 2003), pp.

It was at this time that Muhammad learned very quickly that his claim to leadership among the Quraysh was not going to be well received by some. In particular, certain clans, including the Abdul Dar, refused to hear of it and ridiculed Muhammad's claims. Initially, he was left alone by most of the Quraysh. But when Muhammad clearly stated that all the idols of the Ka'bah were to be expunged, and, only the worship of Allah, the god to whom Muhammad connected his prophetic calling, would be allowed, the other clans' silence turned to ridicule and anger.[13] What Muhammad was essentially demanding was for other clans to surrender the basis of their power and independence and give their submission to Allah—and thus ultimately to Muhammad.

Many of his biographers try to assert that Muhammad was in constant physical danger during this time, but in reality this is greatly exaggerated, as he had the protection of Abu Talib and the Hashim clan.[14] While opponents could ridicule him over his claims, they could not physically harm him. This allowed him the freedom to continue preaching, which included efforts to preach to outlying tribes when they came for the annual pilgrimage. In fact, one Muslim biographer admitted that "Muhammad lived in a free country very much like a republic."[15] This is contrary to the popular perception maintained by some.[16]

But while many refused to heed Muhammad's calls for submission, there were those who did. An analysis of the early male converts to Islam is very revealing, indicating that only nine came from within what could be called the princely clans while thirty-five were outsiders, those from tribes who were excluded from power. And of the former, three were close family to Muhammad, while the other six all came from the Abid Manef line, again invoking parallels to Mao Tse-tung. Not one came from the Abdul Dar, the clan that controlled the other offices of the Ka'bah. Thus, Muhammad's initial appeal was to what could be called the disenfranchised element within Makkah. It is also interesting that his primary moral position throughout his preaching was that the middle class of the Quraysh, that is, those who were becoming independently wealthy from tribal authority, were not giving enough of their wealth away to others.[17] There have been a number who have noted that Muhammad's teaching, and Islam in general, contains a strong dose of socialist philosophy.[18]

98–99; Muhammad Hamidullah, *The Battlefields of the Prophet Muhammad* (repr., New Delhi: Kitabbhavan, 2003), p. 5.

[13] Ibn Sa'd, *Kitab al-Tabaqat al-Kabir,* trans. S. Moinul Haq (New Delhi: Kitabbhavan, n.d.), 1:230–31.

[14] *Al-Tabari,* p. 1174.

[15] Muhammad Haykal, *The Life of Muhammad*, trans. Isma 'il Raji al-Faruqi (Selangor, Malaysia: Islamic Book Trust, 2002), p. 92.

[16] *Muhammad as a Military Leader*, p. 12.

[17] *Muhammad at Mecca*, pp. 70–71.

[18] Karen Armstrong, *Muhammad* (New York: HarperCollins, 1992), p. 92; Maulana Muhammad Ali, *The Religion of Islam* (Dublin, Ohio: Ahmadiyya Anjuman Isha'at Islam Lahore, 1990), p. 509. Also see pp. 346, 510–12.

But Muhammad went beyond direct proclamations by cultivating a fifth column within the elite of Makkah itself. While there is no evidence to say that Abu Talib ever became a Muslim, he certainly used his leadership in the Hashim clan to protect Muhammad from physical harm. But a more important personage was found in Muhammad's uncle al-Abbas bin Abdul Muttalib. Al-Abbas was a man of considerable wealth and position in the Qurayshi elite, being both an active merchant and one who extensively loaned money—essentially a banker. His position also made him a valuable conduit for information, and, throughout Muhammad's campaign against the Quraysh, al-Abbas provided money, supplies, and intelligence to his nephew in support of his operations.[19] Incredibly, his activities never seemed to arouse the suspicions of other leaders within the Quraysh. Indeed, it is possible that others in the Quraysh unknown to us today were also recruited to support Muhammad's activities and thus refused to stop al-Abbas, or even assisted him. An active fifth column within the ranks of the Quraysh would be a constant problem for them until the fall of Makkah in 630 AD.

While violence against converts to Islam was minimal, with only two killed in seven years and even this is in dispute, some of the Quraysh finally lost patience with Muhammad.[20] They first offered Muhammad the title of King of Makkah, but he refused this, as to accept would mean he had received his authority at their hands. Next, they tried a boycott of the Hashim clan, which lasted three years but lacked substance as some within the Quraysh violated the terms and sent supplies to those proscribed. When the boycott collapsed, some of the Qurayshi leaders finally made a halfhearted attempt to murder Muhammad. He and his converts, over a number of months, migrated north to Yathrib, later called Madinah, where Muhammad had relatives who were willing to protect him.[21] Moreover, he had managed to convince two of the main tribes there, the al-Khazraj and al-Aws, to allow him to be the judge over their disputes. In making the Second Pledge of Aqabah, the tribal leaders of these two groups told Muhammad that making the covenant with him would require that they break their treaties with the major Jewish tribes in Madinah.

---

[19] *Battlefields of the Prophet Muhammad,* p. 120; Ibn Ishaq, *The Life of Muhammad*, pp. 301, 309; Abu Ja'far Muhammad bin Jarir al-Tabari, *The History of al-Tabari*, vol. 7*, The Foundation of the Community,* trans. M.V. McDonald (Albany: State University of New York, 1987), pp. 1344–45. Al-Abbas was present at the Battle of Badr, but Muhammad noted that he was there against his will. He ensured his uncle was protected and coolly demanded a ransom from him for his release so as to maintain his cover.

[20] Maulana Muhammad Ali, *Muhammad the Prophet* (Columbus, Ohio: Ahmadiyya Anjuman Isha'at Islam Lahore, 1993), p. 50. This author is one of the few writing on Muhammad to mention these deaths, basing them partly on Ibn Sa'd's *The Women of Madina,* trans. Aisha Bewley (London: Ta-Ha Publishers, 1997), pp. 185–86. Sa'd's account seems apocryphal as it is unlikely Abu Jahl could have gotten away with open murder without inciting some type of blood feud.

[21] The City. The longer title, *al-Madinatu 'n-Nabi*, means The City of the Prophet.

Muhammad concurred and supported this action, information that was apparently kept from the Jewish tribal leaders.[22] He had set himself on the road of armed conflict within the very community to which he was migrating.

With the migration to Madinah in 622, Muhammad now had a relatively secure base of operations. However, this base had internal problems, both with the tribes who supported him as well as with the three major Jewish tribes that refused to enter into a formal treaty with him. Muhammad and his approximately one hundred other emigrants established the Covenant of Madinah, in which he laid out the basic functions of government in the town, the relations of the covenanting tribes to each other, and the fact that none would aid the Quraysh.[23] While minor Jewish clans entered into this agreement, the three major Jewish tribes, the *Banu* Qaynuqa, al-Nadir, and Qurayzah, were not included.[24] This is a critical issue, often neglected or distorted in histories and biographies.

Muhammad set out to organize his base and prepare to engage the Quraysh at their weakest link—their economic livelihood. The mission was, in the words of his closest supporter Abu Bakr, to bring people to Allah and the right path, and "those who deviated from it, they were punished by the Messenger of Allah till they became Muslim."[25] Within Madinah, there were two groups of Muslims: the *muhajirin*, or emigrants from Makkah, and the *ansar*, or helpers who already resided there. With only a few exceptions, the *muhajirin* did not engage in much productive work, although they were apparently supposed to in order to assist the *ansar* in harvesting the latter's date crop.[26] Instead, many of the *muhajirin* resided in the *suffah*, a section of Muhammad's newly built mosque where unemployed families awaited largess as well as recruitment for various *razzias*, or raids.[27] The latter, being a distasteful term from the "days of ignorance," or *jahiliyyah*, was changed to *ghazwah* or *sariyyah*—expeditions.[28]

But expeditions required funding, and so Muhammad instituted the payment of *zakat*, or the charity tax, along with the normal tribal *shaykh's* requirement

[22] Ibn Ishaq, *The Life of Muhammad*, p. 203; *Al-Tabari*, pp. 1220–21.

[23] The number of initial immigrants is vague. This is from a list compiled by Ibn Ishaq, *The Life of Muhammad*, p. 218.

[24] "The Treaty of Madina," in Sultan Ahmed Qureshi, *Letters of the Holy Prophet Muhammad* (repr., New Delhi: Kitabbhavan, 2003), pp. 34–42. *Banu* means "tribe," and is a common title used.

[25] "To all the Rebel Tribes," in Hafiz Muhammad Adil, trans., *Letters of Hadrat Abu Bakr Siddiq* (New Delhi: Kitabbhavan, 1994), pp. 1–2. Al-Tabari's version is somewhat different. See *The History of al-Tabari*, vol. 10, *The Conquest of Arabia,* trans. Fred M. Donner (Albany: State University of New York, 1993), p. 1882.

[26] *Sahih Imam Muslim,* trans. Abdul Hamid Siddiqi (New Delhi: Idara Isha'at-E-Diniyat, 2001), vol. 5, no. 1771, p. 191 [bk. 19, no. 4375].

[27] *Al-Tabaqat,* 2:300–302; *Al-Bukhari,* vol. 5, no. 702; *Muslim,* vol. 8, no. 2769R3, p. 275 [bk. 37, no. 6672]; *Battlefields of the Prophet Muhammad*, p. 136.

[28] A *ghazawah* was one led by the Prophet, while a *sariyyah* was a smaller operation led by a lieutenant.

that of any booty collected, or even mineral wealth mined, one-fifth would go to Muhammad as *khums*.[29] While the first is seemingly innocuous—after all, who is against charity—the tax was necessary to provide badly needed funds to conduct operations. The second helped fill the coffers at critical moments in the early days of the movement and would later accumulate as Muhammad demanded the same from tribal leaders who came and gave their Islam.[30] Financing is one of the most neglected aspects of research in asymmetric warfare, as one can search in vain to discover any discussion regarding the financial aspects of these operations. Apparently in the minds of many analysts, asymmetric soldiers do not need a place to sleep, food to eat, women to soothe their hearts, or entertainment to enliven their souls. The base of operations needed funding, and *zakat* was one of the early principal means.

As the *muhajirin* had no property and little wealth, the burden of *zakat* fell on the *ansar*. Madinah was particularly suited for this as it had its own agricultural base, something lacking in Makkah. While liquid assets were minimal, *zakat* could be paid in kind and thus troops from the *muhajirin* would at least have rations, while any excess could be traded. By using this system, Muhammad got all to participate in expeditions, either directly or indirectly. Yet, while the most important aspect of his operations, the logistical base of the *ansar* was also the most obscure.

It took Muhammad about seven months to position himself to make the first raids against the Quraysh. The area from Makkah northward in western Arabia is called the al-Hijaz and consists of a narrow coastal plain bordering the Red Sea, with a significant rise in elevation culminating in a jagged range of steep mountains and a plateau of lava rock. Zigzagging along the seacoast was the western caravan route between Makkah and Syria, which was the major supply line for the merchants of the Quraysh. The vertical drop from the interior heights to the caravan route is approximately 2,300 feet. Thus, the plateau essentially provided an elevated parallel path to the trade route, offering Muhammad the operational high ground. Numerous mobility corridors of varying size were perpendicular to this route, and this provided Muhammad almost unlimited access to the Qurayshi caravans at any point of his choosing. As to the operational area, the primary zone of initial operations involved close to 100,000 square kilometers, being an area

[29] For an example, see "To the Chiefs of 'Abahila, Hadar Maut," in *Letters of the Holy Prophet*, pp. 131–32. Also, in the early days of Islam, *zakat* was also called *saqadah*. Sura 2, recited soon after Muhammad's arrival in Madinah, contains the earliest references to the "poor rate," or *zakat*. See Sura 2:83. While Muslim writers defend it as a charity tax, it must be recalled that at this time the poor were in the *suffah*. See Abdul Hameed Siddiqui, *The Life of Muhammad* (Jaya, Malaysia: Islamic Book Trust, 1999), pp. 154–56.

[30] "To the Chiefs of 'Abahila, Hadar Maut," p. 119; "Letter to Bani Zuhair;" p. 126; "Injunctions to the Tribe of Lakham;" p. 134; and "To the Chiefs of Yemen." All in *Letters of the Holy Prophet*.

the size of the state of Indiana. Later operations would include an area three to four times that.

Muhammad's early army, if it could even be called that, seemed to be a motley ill-assortment of men who were poorly equipped and lacked experience in field operations. They were particularly hampered by lack of transport, having few camels and even fewer horses. Horses were particularly prized, as they were fast and provided their riders with the advantage of height in battle. Furthermore, the Arabs had already developed the stirrup, making their horse-mounted fighters far more effective than their Byzantium counterparts. But while horses were fast they were constrained by the nature of the desert environment. Lack of water coupled with soft sand or sharp rocky areas limited their use to certain terrain. Camels were not as fast nor considered the best animals to ride into battle, but their large feet being twice that of a horse's lowered the ground pressure they exerted and made them ideal for soft areas. Their legendary endurance through dry areas added to their utility, but like horses, they too were sensitive to rocky areas.

The Quraysh had the most mobile field force in western Arabia, and possibly the entire peninsula. With the money necessary to purchase large quantities of animals, they could field caravans with up to 2,500 camels and an elite cavalry force of several hundred horsemen led by competent field officers like Khalid bin al-Walid. Furthermore, their men could afford armor, spears, and quality swords. Being the most modern and well-equipped army on the field, they could also call upon other tribes to support them as part of a coalition. On the other hand, Muhammad's men had to fight almost exclusively on foot. Only occasionally did Muhammad dispatch his early expeditions with camels, and then the raiders had to take turns riding them. They had difficulty getting decent weapons and initially subsisted on what they could plunder from their enemies. Even food could be scarce on some expeditions.[31]

Nevertheless, Muhammad had some significant advantages over the Quraysh. His army, though small, quickly became elite, motivated by a single driving force—the will of Muhammad and his developing ideology. The Quraysh were divided in their opinions and more concerned with earning wealth, especially during the *hajj,* than they were in engaging in the hardships of field campaigning.[32] While both sides engaged in active intelligence operations, Muhammad was able to expand his fifth column within the Quraysh to keep them divided and confused, while at the same time providing him with unparalleled levels of intelligence. While the Quraysh were nervous about waging war against people who they still considered to be their brethren,

[31] *Al-Bukhari,* vol. 5, no. 647, p. 648.

[32] Ibn Ishaq, *The Life of Muhammad*, pp. 459–60; Al-Tabari, *The History of al-Tabari,* vol. 8, *The Victory of Islam,* trans. Michael Fishbein (Albany: State University of New York, 1997), p. 1884; *Al-Tabaqat*, 2:85. The references deal with the siege of Madinah by the Quraysh, and the siege was conducted two months before the annual *hajj*. It was necessary for the Quraysh to return to provide resources for the expected pilgrims.

Muhammad had stripped the Muslims of their family ties and identity, uniting the *muhajirin* and *ansar* into one new tribal brotherhood, with more concern for obedience to the Prophet than loyalty to traditional family, clan, or tribe, thus establishing a deep psychological cohesion.[33] Finally, the Quraysh were actually squeamish, with a few exceptions, about the horrors of the battlefield. In contrast, Muhammad and his men quickly gained an incredible and even insatiable love for the spilling of human blood and gore, even that of their own.[34]

An analysis of Muhammad's offensive campaign against the Quraysh reveals five major phases of operations. The first, from 622–623, focused on training raw recruits, developing tactics, and establishing some local tribal alliances to start isolating the Quraysh. The second, from 624–625, saw the first early successes, including victory at the first pitched battle between the two sides at Badr in 624, but also included the first major setback with the defeat of Muhammad's force at the base of Mount Uhud, north of Madinah, in 625. During this second phase, Muhammad also began his campaign to consolidate his hold on Madinah, which meant the expulsion of the *Banu* Qaynuqa and al-Nadir, two of the rival Jewish tribes, and the seizure of their property and wealth. He had originally intended to kill the men of the *Banu* Qaynuqa and al-Nadir but had been constrained by various appeals. On a later date, the *Banu* Qurayzah would not be so fortunate, as Muhammad's control over the town tightened.

The third phase, from 625–627, saw Muhammad regrouping his forces, developing more tribal alliances, often by raiding his neighbors with such ferocity that they would virtually beg to give their Islam, or barring such, despoiling them until they reached a desperate state of affairs.[35] In one year of raids, the Muslims seized at least 700 camels and 2,000 goats, plus an undefined amount of cattle, with virtually no property loss to themselves.[36] They also continued their raids on the Qurayshi caravans and held back a Qurayshi counteroffensive at Madinah in what turned out to be the high tide of the enemy opposition. In this latter operation, called the Battle of the Trench,

---

[33] The consequences of such cohesion were deep. One Muslim indicated that if so ordered, he would kill his own brother, Ibn Ishaq, *The Life of Muhammad*, p. 369. Another offered to kill his own father. See ibid., p. 492.

[34] Prowess in combat was important to the Arab people, but killing was normally avoided in raids. Examples of this desire to kill is seen in Muslim poetry, though some Quraysh began to respond in kind. See Ibn Ishaq, *The Life of Muhammad*, pp. 341, 347, 348, 357, 366, 369, 408–09, 411. While most Quraysh lamented death, Muslim poets often celebrated it, extolling the abilities of martyrs like Muhammad's uncle, Hamzah. See poetry in ibid., pp. 419–21. An example of one reciting explicit poetry of how he would die was Abdullah bin Rawaha, prior to the Battle of Mu'tah. See ibid., p. 532. There are also numerous examples of Muslims killing with little or no provocation, ibid., pp. 447, 673–75.

[35] *Al-Tabaqat*, 2:108–09.

[36] Ibid., 2:97–117. The numbers given are only of that listed. Other raids indicated livestock captured but provided no quantities.

the Quraysh had assembled a coalition of other tribes that allowed them to field an army of 10,000 men, with at least 300 cavalry.[37] Using deception and deceit, Muhammad was able to divide the coalition, while engaging in a month-long battle of attrition using tactics the Quraysh had never before seen.[38] Despite numerical and technological superiority, the Quraysh withdrew from Madinah, with Muhammad noting that the Muslims would now raid them mercilessly.[39] He was right. But before completing his dominance of the Quraysh, Muhammad removed the last obstacle to his complete control of Madinah, that being the slaughter of the warriors of the *Banu* Qurayzah.

The destruction of the *Banu* Qurayzah served as an object lesson the likes of which many tribes in the area had never seen. While Muhammad had ordered the dismemberment of men who stole his camels, or the violent attack on a tribe who raided his caravans,[40] it was the beheading of about 800 warriors of this last Jewish tribe that gave notice to all around that Muhammad would brook no opposition. While much is made by biographers that the *Banu* Qurayzah violated a treaty with Muhammad, no such treaty has ever been produced, nor is there any real evidence such existed. In fact, one *hadith,* or tradition, indicates that at best there was a "non-binding" agreement between the two sides and thus no formal treaty at all.[41] Even the Qur'an implies that Muhammad could violate a peace simply because he suspected a tribe might prove treacherous.[42] When one recalls that Muhammad had already agreed with the other Madinah tribal leaders at the Second Pledge of Aqaba that they would violate their treaties with the main Jewish tribes and, moreover, that Muhammad had already previously announced openly to the Jews that he intended to expel them from Madinah, it comes as no surprise that this final confrontation would occur.[43] Evidence within Islamic sources indicates that Muhammad took measures to provoke such conflicts, and then used minor incidents to justify the suppression or destruction of his enemies.

With Madinah now firmly in his control, Muhammad stepped up his operations, boldly advancing on Makkah in 627. This time, he had overplayed his hand. The Quraysh, fearing an armed Muslim force moving into their town, met Muhammad just outside of Makkah and forced him to consent to the Treaty of al-Hudaybiah. Its terms were clearly advantageous to the Quraysh, so much so that it caused serious dissension within the ranks of Muhammad's army.

---

[37] Ibid., 2:80–81.

[38] Ibn Ishaq, *The Life of Muhammad*, p. 454.

[39] *Al-Bukhari,* vol. 5, no. 435.

[40] Ibid., vol. 4, no. 261A; vol. 5, no. 505. However, the *Al-Tabaqat* indicates it was a case of *lex talionus.* See 2:114–15. However, the other *hadith* does not mention this. *Al-Tabaqat*, 2:111–12.

[41] Ibid., 2:95.

[42] Sura 8:58; *Tafsir Ibn Kathir*, 4:341–43.

[43] *Al-Bukhari*, vol. 9, nos. 77, 447.

Sullen, the men retraced their steps back to Madinah, after Muhammad boldly took measures, including the issuance of poetic propaganda, to transform this obvious defeat into an incredible victory.[44]

Entering the fourth phase of operations, from 627 to 630, Muhammad broadened his support base around Madinah. While not gaining many tribes to back him, he did obtain their neutrality, thus further increasing the isolation of the Quraysh. He then embarked on violating the Treaty of al-Hudaybiah, a fact ignored by virtually every one of his biographers. He first secretly supported a proxy who took to raiding the Qurayshi caravans along the western route, followed by his support for a group of women who migrated to Madinah.[45] Finally, he returned to Makkah the following year for a pilgrimage, fully armed in violation of the treaty's seventh provision.[46] While the Quraysh protested these actions, they did virtually nothing to enforce the treaty. By their inaction, they had demonstrated their weakness. In January 630, now with a force of 10,000 men, Muhammad made his triumphal march on Makkah. The Qurayshi resistance decisively collapsed aided by the treachery of some of their own leaders, and Muhammad occupied their city. The only casualties were a handful of people put to death for insulting him or backing out of their Islam.[47]

Phase five, from 630 to 632, was the final consolidation of victory. A number of campaigns were conducted, some ranging as far north as Syria, where an outnumbered Muslim force was compelled to retreat before a large Byzantine army. Other campaigns drove reluctant tribes into the Muslim fold, and Muhammad spent a great deal of time receiving the Islam of many distant tribal leaders and writing what amounted to ultimatums to others, including ones to Byzantium and the Sassanid.[48]

Muhammad's campaign against the Quraysh was in large measure an asymmetric one, in which he generally refused to engage his enemies on their terms. He used tactics and principles quite foreign to both the enemies of his time and to Westerners of today. These principles are in large measure very similar to those used by radical political movements of the 1960s, which were summed up in Saul Alinsky's book *Rules for Radicals*.[49] These principles could very well form the basis for a paradigm shift in the way the United States, and the West in general, views asymmetric warfare, in particular involving Islam (*Table*). It is worth analyzing a few of these principles.

---

[44] Sura 48, *The Victory*; *Muslim*, vol. 5, no. 1785, p. 205 [bk. 19, no. 4405]; *Al-Tabaqat*, 2:122; Ibn Ishaq, *The Life of Muhammad*, p. 507.

[45] Ibn Ishaq, *The Life of Muhammad*, pp. 507–08, 509–10; *Al-Tabari*, pp. 1552, 1553.

[46] *Al-Tabaqat*, 2:150; *Al-Tabari*, p. 1597; "Treaty of Hodaibiya," in *Letters of the Holy Prophet Muhammad*, pp. 46–47.

[47] Ibn Ishaq, *The Life of Muhammad*, pp. 550–51.

[48] "Letter to Heracles Caesar," p. 65, and "Letter to Khusro Perwez, Emperor of Fars," p. 70, both in *Letters of the Holy Prophet Muhammad*.

[49] Saul Alinsky, *Rules for Radicals: A Pragmatic Primer for Realistic Radicals* (New York: Vintage Books, 1972), pp. 126–38.

Table—Rules for Radicals

1. Power is not only what you have, but what the enemy thinks you have.
2. Never go outside the experience of your people.
3. Wherever possible go outside the experience of the enemy.
4. Make the enemy live up to their own book of rules.
5. Ridicule is man's most potent weapon.
6. A good tactic is one your people enjoy.
7. A tactic that drags on too long is a drag.
8. Keep the pressure on.
9. The threat is usually more terrifying than the thing itself.
10. The major premise for tactics is the development of operations that will maintain constant pressure upon the opposition.
11. If you push a negative hard and deep enough it will break through into its counterside.
12. The price of a successful attack is a constructive alternative.
13. Pick the target, freeze it, personalize it, and polarize it.

Source: Saul Alinsky, *Rules for Radicals: A Pragmatic Primer for Realistic Radicals* (New York: Vintage Books, 1972), pp. 126–38.

While symmetric warfare involves direct combative action, asymmetric warfare often involves the mere threat of action. In this case, the principle that "the threat is worse than the event itself" is very applicable. Muhammad created this constant image of the pending threat and used it to full advantage. He did this by selectively using terror to leverage future enemies into submission, a method to which he himself admitted.[50] Such terror soon became legendary, and yet Muhammad could be quite magnanimous to submissive enemies when he so chose. In one case, a group of men from the northern tribe of Urayna came to the Prophet and professed their Islam. They then managed to raid Muhammad's camels and kill the man tending them. Muhammad sent a party in pursuit, and, when they were caught, it was essential for the sake of discipline within the Muslim ranks for him to make an example of them. He had their arms and legs torn from their bodies, and their eyes gouged out with red-hot nails. Screaming in agony, they were then left to die in the midday sun.[51] While one could argue that they deserved to die for murder, the extreme means of execution created a sense of terror and awe among both the faithful and enemies.

A reputation for terror and cruelty preceded his later operations, as witnessed by the cry of despair from the date farmers of Khaybar when they

[50] *Al-Bukhari*, vol. 4, no. 220; *Muslim,* vol. 2, no. 523 [bk. 4, no. 1062].

[51] *Al-Bukhari*, vol. 4, no. 261A, and vol. 5, no. 505.

went to their orchards one morning to find Muhammad's army closing in.[52] But this threat of terror could also be mitigated by compassion and mercy. After the conquest-of-Khaybar operation, a Jewish woman attempted to poison the Prophet. While one of his companions died, Muhammad spit the tainted meat from his mouth and then questioned the woman about her actions. Upon her confession that Muhammad was surely a prophet, and thus strengthening his standing with his people, he had the woman released.[53] Fear and terror would be followed by sighs of relief. Others would experience this same tension that would drive them to give their Islam, as witnessed by the letters sent by Muhammad to neighboring tribal leaders.

Another principle highlighted by Alinsky is the notion that "one should make his enemies live up to their own book of rules." While many biographers are very adulatory of Muhammad regarding treaties, the source materials on his life indicate that he had little trouble with violating agreements, while making it appear as if his enemies were the first to do so. In the early days of his tenure in Madinah, virtually every testimony regarding a *ghazwah* or *sariyyah* that was sent out was preceded by the statement that a report had come to Muhammad that a neighboring tribe was preparing to attack the Muslims. Yet when the army sent to break up this enemy arrived, they would find no enemy army present. But they certainly found plenty of camels, sheep, and cattle, which they promptly seized and brought back to Madinah. Of interest is the fact that after the victory at the Battle of the Trench in 627, there were virtually no more "rumors" of neighboring tribes about to attack. Instead, Muhammad simply sent out his raiding parties to terrorize and pillage those around him.[54] By blaming others, he could justify his own raids upon his neighbors, raids that often went beyond the norms of custom in Arabia.

While it was well known in Arabia that one did not wish to cause fatalities during raids and thus incur a blood feud, Muhammad grew increasingly unconcerned with this custom, though he demanded that his enemies retain such ideals. He only paid the bloodwit, the payment made to make compensation for those killed from other tribes, when it suited his purposes. On a number of occasions, his men committed assassinations or even outright murder, and it was known that Muhammad's men were involved.[55] Yet no effort was made to pay a bloodwit for the deceased. On the other hand, when one of his men killed two men who had just been granted protection by Muhammad, a fact to which the killer was unaware, Muhammad determined to pay bloodwit,

[52] Ibid., vol. 5, no. 510.

[53] Ibid., vol. 4, no. 394; Ibn Ishaq, *The Life of Muhammad*, p. 516; *Al-Tabari*, pp. 1583–84. But in the *Al-Tabaqat*, 2:133, Ibn Sa'd indicates the woman was actually put to death. The weight of evidence favors her pardon.

[54] *Al-Tabaqat*, 2:96.

[55] Ibid., 2:39.

using the incident to create a controversy with the *Banu* al-Nadir to expel them from Madinah.[56]

Another incident well illustrates this principle. Just prior to the conquest of Makkah, Muhammad sent a diplomatic envoy to Heracles, the king of Byzantium. On the envoy's return, he was ambushed and robbed by a raiding party of the clan of Judham. Nobody was injured in this raid, and, a neighboring tribe, the *Banu* al-Dubayb, having heard of this incident, caught up to the Judham raiders and forced them to return the stolen items to Muhammad's envoy. This did not stop Muhammad from sending out his adopted son Zayd with 500 men to raid the Judham, where they killed the clan's leaders along with a number of other men and seized 1,000 camels, 5,000 goats, and took 100 women and children captive. It was later revealed that the Judḥam had given their Islam earlier, and, though the goods and captives were returned, no bloodwit was paid for the dead.[57]

A third principle of Alinsky's worth analyzing is the concept that "one should go outside the experience of their opponent." An important aspect of asymmetric operations is to engage in tactics within the experience of one's own personnel. On the other hand, it is equally important to hit the enemy in ways he does not anticipate nor can counter. During Muhammad's consolidation of his authority in Madinah, he ordered a number of high-profile assassinations. While this may seem common enough in the context of history, in reality it was quite novel for Arabia. A few examples will suffice.

In March 624, he ordered the assassination of Asma bint Marwan, a woman who was reciting poetry against Muhammad. In Arabia, poetry was a form of journalism, and good poets were considered ideal repositories of historical information regarding the region and its tribes. During the night, an assassin infiltrated her home, moved undetected through a room with four sleeping children, removed a baby that was asleep on her breast, and drove a sword through her body.[58]

A few months later, in September 624, Muhammad organized the assassination of a prominent leader of the Jewish *Banu* al-Nadir, Ka'b bin al-Ashraf. He was one of those wealthy middle-class individuals Muhammad had been speaking against and had been reciting poetry and talking to the Quraysh to encourage their resistance to Muhammad after their stunning loss at Badr. With Muhammad's permission, the assassins prepared a ruse to lure Ka'b from his fortified home. Several relatives of the Jewish leader came to offer

---

[56] Ibn Ishaq, *The Life of Muhammad*, pp. 434–37; *Al-Tabaqat*, 2:68–69.

[57] *Al-Tabaqat*, 2:108–09.

[58] Ibn Ishaq, *The Life of Muhammad*, p. 675; *Al-Tabaqat*, 2:30–31. Some have tried to say this incident was fabricated, but many Muslim writers accept it as true, as do pro-Muslim Western writers. See Haykal, *Life of Muhammad*, p. 243; Shawqi Abu Khalil, *Atlas on the Prophet's Biography* (Riyadh, Saudi Arabia: Darussalam, 2003), p. 123; Sir John Glubb, *The Life and Times of Muhammad* (New York: Cooper Square Press, 2001), p. 195; W. Montgomery Watt, *Muhammad at Medina* (Oxford, UK: Oxford University Press, 2004), p. 178.

feigned complaints about the Prophet's rule of Madinah, managed to entice Ka'b to follow them in the darkness, and then cut him down in the road with a dagger thrust. Though the *Banu* al-Nadir knew that Muslims had killed him, Muhammad offered no bloodwit to pay for the crime.[59] Moreover, the day after the assassination, the Prophet issued a proclamation to "kill any Jew that falls into your hands."[60] After that declaration, a Muslim killed a Jewish merchant, and no bloodwit was offered.

Lastly, in the summer of 625, Muhammad heard that a tribe was preparing to attack the Muslims, under the leadership of Sufyan bin Khalid. He assigned Abd Allah bin Unays to assassinate this tribal leader. Abd Allah, after receiving a basic description of the man he was to kill, simply rode into the man's camp. He walked up to Sufyan bin Khalid, introduced himself, and asked him if he was preparing such a raid. Upon confirmation, Abd Allah spoke with him at length, almost as a friend, before finally killing him in his tent and severing his head.[61]

What these assassinations reveal is the incredible lack of security these individuals had, especially the latter who was the leader of a tribe. This indicates that few really took seriously the threat of assassination in those days, and this made them very vulnerable to the tactic. Moreover, the tactic was considered quite despicable by those who survived such attacks and had to bury their friends and loved ones. By using assassination as a tactic, Muhammad moved outside of the experience of his opponents.

This issue of assassinations brings up possibly the most important aspect of asymmetric warfare—the lack of rules. In symmetric warfare, combat is governed by a series of rules and protocols today enshrined in a series of international conventions, which were the culmination of centuries of experience in the Western world. Influenced by the Christian worldview of the Western church, the laws of warfare were intended to protect noncombatants and alleviate the suffering of those who could not defend themselves. The problem with asymmetric warfare is that there are no rules. This is what makes fighting an asymmetric enemy so difficult. The principles of warfare used by Muhammad broke all conventions and rules of his day. He still demanded that his enemies adhere to these rules, while he violated them whenever convenient for his objectives.

The fall of Makkah in 630 did not bring an end to the Quraysh. Instead, Muhammad stripped the middle class of its privileges and a principal means of increasing wealth, the use of interest, or *riba*. Concurrently, he reconfirmed the position of the Qurayshi leaders in controlling the Ka'bah and any money collected for it.[62] But instead of the burden of financial support falling on the Quraysh, it was now to be born by all. While the princely class still retained

[59] Ibn Ishaq, *The Life of Muhammad*, p. 368; *Al-Tabaqat*, 2:39.
[60] *Al-Tabari*, pp. 1372–73.
[61] *Al-Tabaqat*, 2:60.
[62] Ibn Ishaq, *The Life of Muhammad*, pp. 554, 557; *Al-Tabari*, p. 1642.

its power, the rest of Arabia was now required to financially support these people in their position through taxes like *zakat* and *ushr*, and the threat of middle-class competition for leadership had been stymied. The rule of the Quraysh was now so firmly established that when Muhammad died, leaders within their ranks, including the son of one of Muhammad's key opponents, would rule as the *Khalifa* and run the Islamic state in its early years of foreign conquest. In fact, Abu Bakr would appeal to the *ansars* that it was his right to rule as the first *Khalifa* as the Quraysh were born for such.[63]

The establishment of Islam as the ruling element in Arabia also stripped the tribal leaders of any real power. Power and authority were now concentrated in the hands of the Prophet and his companions, and tribal leaders were to submit major decisions to them. This loss of independence meant unification in Arabia and thus served to provide the political and financial basis for the early triumphs of Islamic conquests throughout the Middle East. Moreover, as the neighboring tribes gave their Islam, other Muslims could no longer raid and plunder them. After Muhammad's death, the *khalifas* had to look to greener pastures, which boded ill for their non-Islamic neighbors to the northwest and northeast.[64]

Lastly, one finds a fascinating and very contemporary comparison between Muhammad and the early Muslims and their primary opposition of the Quraysh. While the latter was technologically and numerically superior, they lacked the drive and determination to kill their enemy, and there are also indications that they did not believe in ultimate victory.[65] In fact, it is difficult to determine if they even understood that the early Muslims were their true enemies. More concerned about economic prosperity and the opinions of other tribes in the area, the Qurayshi leaders were unwilling to implement the measures needed to deal the fatal blows to Muhammad's insurgency. Equally important was the infiltration within their own ranks of secret Muslims, who not only confused the counsel of the Qurayshi leadership, but also provided valuable intelligence and financing to Muhammad, and even helped protect the property of émigré Muslims within Makkah.[66] Indecisive and confused, the Quraysh were unable to maintain their coalitions, thus finding themselves increasingly isolated. When Muhammad's army marched on Makkah, the Qurayshi leaders all but collapsed with their principal men rushing to give their Islam.

In contrast, Muhammad and his men were almost ruthless to the point of excess. They used methods and tactics that stunned the genteel sensibilities of the Quraysh and even the neighboring tribes, such as assassinations and brutal killings as punishments for what were often minor offenses within the

---

[63] *Al-Tabari,* pp. 1840–42. The *Khalifa* in question was Mu'awiyah bin Abu Sufyan (661–680), and one of the key founders of the Umayyad dynasty.

[64] Efraim Karsh, *Islamic Imperialism* (New Haven, Conn.: Yale University Press, 2006), p. 19.

[65] *Tafsir Ibn Kathir,* 4:274; *Al-Bukhari*, vol. 4, no. 276, and vol. 5, no. 375.

[66] *Al-Bukhari,* vol. 4, no. 314.

context of the Arabian world. With the surrender of the mighty Quraysh, neighboring tribes swiftly gave their Islam to the new ruler of the Arabs and paid their tribute accordingly. But while the killings could be brutal, they were selective and limited, thus serving as object lessons for those who were hesitant to submit.[67] The mere threat of a Muslim assassin's blade or massed beheadings was now enough to get a recalcitrant tribal leader to tender his submission and pay *zakat* and *khums* to the Prophet. Indeed, Muhammad's selectivity in choosing his object lessons meant that only a few thousand died during his ten years of campaigning. This alone marks him as one of the most remarkable practitioners of asymmetric warfare.

[67] Sura 8:57; *Tafsir Ibn Kathir*, 4:341.

Southern (Dis)Comfort: British Phase IV Operations in South Carolina and Georgia, May–September 1780

Steven J. Rauch

On 8 June 1780, Lt. Col. Thomas Brown, commander of the Loyalist provincial King's Carolina Rangers, led his regiment into the frontier town of Augusta, Georgia; the place where his Whig neighbors had once tormented, tortured, and disfigured him.[1] In August 1775, Brown had obstinately supported the British Crown against the Whigs and their traitorous actions. For holding those convictions, he was confronted by a large mob; hit in the head with a rifle, which fractured his skull; tied to a tree; and had burning pieces of wood stuck under his feet. Next, his enemies scalped the hair from his head in three or four places. He lost two toes due to the burning he suffered when he was tarred and feathered. Brown was then paraded through Augusta in a cart while ridicule and insults were heaped upon him by the Whig "Patriots."[2] "Burnt Foot" Brown, as he came to be derisively called, spent the next several years leading a regiment of Loyalist rangers in Florida and Georgia, keeping alive the hope that Britain would overthrow the radical regimes of the Whigs and restore peaceful government to the region.

The successful British capture of Charleston on 12 May 1780 had made Brown's triumphal return to Augusta possible. There, American Maj. Gen. Benjamin Lincoln surrendered almost six thousand Continental and militia troops to a powerful British joint land and naval force commanded by Lt. Gen. Henry Clinton.[3] From the British perspective, it was "mission accomplished," as the most important city in the southern states fell into their hands. It seemed all that remained were minor postcombat operations to destroy an inconsequential number of fanatical Whig "dead-enders." By the end of June 1780, columns of British and Loyalist troops, such as Brown's, had overrun South

---

[1] For this paper, I will use the terms Whigs and Loyalists to describe Americans who held different political views. I avoid the use of the words "Patriot" or "American," as they can apply to either side; both were native, both were patriotic, and both were American. Political ideology served as the mechanism for division.

[2] The definitive book on Brown is Edward J. Cashin, *The King's Ranger: Thomas Brown and the American Revolution on the Southern Frontier* (Athens: University of Georgia Press, 1989).

[3] The most recent account of the campaign for Charleston is found in Carl P. Borick, *A Gallant Defense: The Siege of Charleston, 1780* (Columbia: University of South Carolina Press, 2003). For the number of prisoners and casualties of the siege of Charleston, see p. 222.

Carolina and Georgia, garrisoned major population centers, established forward operating bases from the coastal towns deep into the backcountry piedmont, and begun a comprehensive program to organize, train, and equip units of native Loyalists to help restore order and stability to the region. However, subsequent events demonstrated that British forces operated in a complex social, economic, and military environment, one in which commanders soon found themselves struggling to hold fixed bases against attack, protecting supply convoys from ambush, searching for and fighting bands of insurgents inspired by regional leaders, and trying to coexist with a populace where friend was often indistinguishable from foe.

The purpose of this paper is to explore some of the challenges faced by the British army of liberation—or occupation, depending on your view—during the immediate months that followed what appeared to be a conflict-ending, decisive victory at Charleston. The British faced a new phase of the southern campaign, one which we identify today as Phase IV operations or postconflict operations. Phase IV operations include those tasks designed to build a secure and stable environment so that political, economic, and social reconstruction can occur.[4] One of the greatest challenges during this phase is effective management of the transition through policies designed to promote a unified effort and mitigate any attempts at disruption by determined and often fanatical opposition. Recently, historian Joseph Ellis framed this challenge in the form of a historically enduring question: "Can a powerful army sustain control over a widely dispersed population that contains a militant minority prepared to resist subjugation at any cost?"[5] By using the example and experiences of Loyalist commander Brown, I hope to illustrate some of the challenges of occupation that he and other British commanders faced as they sought to restore peace, stability, and security to Georgia and South Carolina during the summer of 1780.

Operation Southern Campaign I (1775–1778)

During 1775–1776, southern Whig leaders and their supporters overthrew the royal governments of the southern colonies of Virginia, North Carolina, South Carolina, and Georgia in relatively bloodless coup d'états. The southern Whigs were indirectly aided in their efforts when the British chose to focus most

[4] The phases of an operation include Phase I (preparation), Phase II (initial operations), Phase III (combat operations), and Phase IV (postcombat operations). The topic of Phase IV operations related to the war in Iraq has generated several discussions in recent literature. See Thomas E. Ricks, "Army Historian Cites Lack of Postwar Plan," *Washington Post*, 25 Dec 2004, p. A01; Col. Kevin C. M. Benson, "OIF Phase IV: A Planner's Reply to Brigadier Aylwin-Foster," *Military Review* (March–April 2006): 61–68; Lt. Col. (Ret.) Conrad C. Crane, "Phase IV Operations: Where Wars Are Really Won," *Military Review* (May–June 2005): 27–36; and Andrew Garfield, *Succeeding in Phase IV: British Perspectives on the US Effort to Stabilize and Reconstruct Iraq* (Foreign Policy Research Institute, September 2006).

[5] Joseph J. Ellis, "Washington: The Crying Game," *Los Angeles Times*, 29 Dec 2006.

of their military power on containing and suppressing the Whig rebellion in the northern colonies. However, the British did undertake a short and inconclusive raid on Charleston in June 1776 by land forces commanded by Clinton and naval units directed by Admiral Peter Parker. The defense of Charleston, enabled by a fort of palmetto logs on Sullivan's Island, provided the Whigs a physical and moral victory over British conventional military forces.[6] The 1776 raid on Charleston was to have been timed with an uprising of North Carolina Scotch-Irish Loyalists, who had expected earlier assistance and were prematurely defeated by the Whigs at the battle of Moore's Creek Bridge in February 1776.[7] Coincidently with the Charleston raid, open warfare broke out between Cherokee Indians and backcountry settlers in Georgia, South Carolina, and North Carolina, which appeared to validate an open alliance of the British and Indians against the backcountry people.[8]

After the victories over the British, Loyalists, and Indians in 1775–1776, the southern Whigs enjoyed almost two and a half years of relative peace. They used this breathing space to organize their governments to include absorbing the existing militia structures, tax structures, legal systems, and legislative assemblies. Between 1776 and 1778, the Whig governments conducted several limited military campaigns against pockets of Loyalists, the Creek and Cherokee, and the loyal British colony of East Florida. These campaigns exercised the militia systems and provided valuable experience for a small cadre of Whig leaders. It was also during this time that southern Loyalists were subjected to varying degrees of repression, to include murder, and many Americans, such as Thomas Brown, found themselves facing the choices of compromising their values so they could retain their property, fleeing their homes to more stable parts of the empire, or taking up arms and fighting against the rebellion.[9] By the end of 1778, an uneasy détente

---

[6] The palmetto tree on the South Carolina state flag commemorates this event.

[7] For a good overview, see David K. Wilson, *The Southern Strategy: Britain's Conquest of South Carolina and Georgia, 1775–1780* (Columbia: University of South Carolina Press: 2005), pp. 5–58.

[8] For information on the early campaigns, see John W. Gordon, *South Carolina and the American Revolution: A Battlefield History* (Columbia: University of South Carolina Press, 2003), pp. 34–57. See also Jeff Dennis, "Southern Campaigns Against the Cherokees: A Brief Compilation," *Southern Campaigns of the American Revolution* 2, no. 10 (October 2005): 17–19, available at www.southerncampaign.org; and Jeff Dennis, "Native Americans and the Southern Revolution, Part II," *Southern Campaigns of the American Revolution* 4, no. 3 (July–September 2007): 21–27, available at www.southerncampaign.org.

[9] On 16 September 1777, the Georgia Whig legislature or General Assembly issued *An Act for the Expulsion of the Internal Enemies of this State* to set up twelve-man committees in each county with authority to judge the political allegiance of any male over the age of twenty-one. Two witnesses were required to verify that he was a "Friend of Freedom," and the man had to swear an oath of allegiance to the state of Georgia. If he was not a Friend of Freedom, he was to be banished from the state under penalty of death, never to return, and the loss of one-half of his property. Heard Robertson, ed., "Georgia's Banishment and Expulsion Act of September 16, 1777," *Georgia Historical Quarterly* (Summer 1971): 274–82.

came to exist between the Whigs and those Loyalists who chose to remain in their homeland.[10]

Change of Course—Operation Southern Campaign II (Georgia)

By 1778, Lord George Germain, secretary of state for the American department, was frustrated by the lack of progress in the war, particularly the defeat at Saratoga and the inconclusive campaigns near Philadelphia in 1777. An important concern was that British military power was limited, and, when France openly became an American ally, the war changed from a regional conflict into a world war.[11] This development stretched British naval and land resources as the empire tried to operate in several theaters of operations, particularly the valuable sugar-producing West Indies.[12] The shortage of manpower had led the British to contract with various German princes for military forces, commonly known as Hessians, to augment the small regular British Army conducting operations in America.

Thus, by 1778, the British government sought to change its strategy, and it appeared that operations in the southern colonies could achieve success. The "soft underbelly" held sparsely settled populations, a large slave population that might be exploited, and valuable export economies based on rice, indigo, beef, hides, and naval stores.[13] On 8 March 1778, Germain ordered an expedition be sent to recover Georgia and the Carolinas. In general, the British adopted a two-phased approach to recover the South. They based the first phase on a swift military invasion to destroy or capture Whig combat units and the targeting of Whig leaders for capture and removal from power and influence. In the second and more complex phase, they would restore peace through reconstituted royal civil governments.[14] Human intelligence analysts classified the southern populace as generally Loyalists who would enthusiastically embrace the opportunity to overthrow their tyrannical Whig governments when British military power appeared.[15]

---

[10] For events of this period, see Kenneth Coleman, *The American Revolution in Georgia, 1763–1789* (Athens: University of Georgia Press: 1958), and Cashin, *King's Ranger*.

[11] The best account of the British perspective of the war remains Piers Mackesy, *The War for America, 1775–1783* (Cambridge, Mass.: Harvard University Press, 1964); Bison book ed. (Lincoln: University of Nebraska Press, 1993).

[12] Mackesy, *War for America*, p. xv.

[13] Ibid., p. 159.

[14] Alan S. Brown, ed., "James Simpson's Reports on the Carolina Loyalists, 1779–1780," *Journal of Southern History* 21 (November 1955): 513; Mackesy, *War for America*, p. 233; Wilson, *Southern Strategy*, pp. 59–64; Gordon, *South Carolina*, pp. 58–62; and John S. Pancake, *This Destructive War: The British Campaign in the Carolinas, 1780–1782* (Tuscaloosa: University of Alabama Press: 2003), pp. 1–19.

[15] The question of how many Loyalists there were among the American population has been addressed in several sources. The most useful is Paul H. Smith, "The American Loyalists: Notes on Their Organization and Numerical Strength," *William and Mary Quarterly* 24 (April 1968): 259–77. Smith concludes that about 19,000 men served in provincial or Loyalist militia units

The constrained force structure drove British planners to use the minimum number of "boots on the ground" and to rely on Loyalists to assume a significant role in their own liberation, to include overthrowing, capturing, and detaining former Rebel leaders. As John Shy has described this effort, "No longer would British troops try to occupy and hold directly every square foot of territory; instead, the war was to be 'Americanized'—territory once liberated would be turned over as quickly as possible to loyal Americans for police and defense, freeing redcoats to move on to the liberation of other areas."[16] These Loyalist security forces would then assist in the reestablishment of civil government, the economy, judicial functions, and public safety.

The successful invasion of Georgia in December 1778 by forces led by Lt. Col. Archibald Campbell and Maj. Gen. Augustine Prevost seemed to validate all the expectations of the new strategy. The defeat of the Georgia Whig regime was hailed as a success, and many exiled Georgia Loyalists returned to the homes they had fled earlier. However, there were indications that Whig resistance would be offered from South and North Carolina. In February 1779, a combined force of Georgia and South Carolina Whigs destroyed a South Carolina Loyalist force at Kettle Creek, Georgia. That incident was perhaps overshadowed by the stinging defeat in March 1779 of Georgia, North Carolina, and South Carolina militia and Continental forces at Briar Creek by British forces.[17] The British determined that the backcountry region near Augusta and beyond was too difficult to control and instead appeared satisfied with a foothold at Savannah and along a thin corridor of the Georgia coast. Occupation was not stagnant; during the spring, Prevost attempted another inconclusive raid on Charleston and, more dramatically, defended Savannah against the joint American and French expedition in October 1779. To anyone paying attention, the operations in Georgia revealed an early indication of the complex problems facing an occupying military force attempting to pacify a colony rent by the stresses of civil war and rebellion against authority.[18]

---

in all areas throughout the war. From that figure, he concludes that there were about 128,000 adult male Loyalists out of a total Loyalist population of 513,000 men, women, and children, or 19.8 percent of white Americans.

[16] John Shy, "British Strategy for Pacifying the Southern Colonies, 1778–1781," in Jeffrey J. Crow and Larry E. Tise, *The Southern Experience in the American Revolution* (Chapel Hill: University of North Carolina Press, 1978), p. 159.

[17] For good recent summaries of these battles, see Wilson, *Southern Strategy*, pp. 65–100.

[18] Martha Condray Searcy, "1779: The First Year of the British Occupation of Georgia," *Georgia Historical Quarterly*, 63 (Summer 1983): 168–88; Kenneth Coleman, "Restored Colonial Georgia, 1779–1782," *Georgia Historical Quarterly* 40 (March 1956): 1–20; and Patrick J. Furlong, "Civilian-Military Conflict and the Restoration of the Royal Province of Georgia, 1778–1782," *Journal of Southern History* 38 (1972): 415–42.

Operation Southern Campaign III (South Carolina)

In March 1779, Germain advised Clinton that he should follow up success in Georgia with a more decisive operation to recover South Carolina.[19] To help obtain intelligence about the inhabitants' attitudes toward such an operation, Germain ordered James Simpson, former royal attorney general for South Carolina, to Georgia and South Carolina to report on conditions. On 28 August 1779, Simpson reported to Germain and Clinton that the Loyalists of both Georgia and South Carolina had been relentlessly persecuted since 1776 and anxiously sought assistance to overthrow their Whig oppressors. Simpson stated,

> Unless the government was to be so firmly established as to give security to them without protection of the Army . . . the success would be far from complete. And if upon a future emergency, the Troops were withdrawn . . . they should suffer. I am of the opinion whenever the King's Troops move to Carolina they will be assisted by very considerable numbers of the inhabitants. . . . If the terror [the Whigs] have excited was once removed, a few months would restore this country to its former good government."[20]

All information seemed to indicate that a significant number of Loyalists were ready to assist the British military with manpower and political support.

Clinton departed New York City on 26 December 1779 with 8,708 soldiers aboard a fleet of eighty-eight transports accompanied by thirty warships.[21] V. Adm. Marriot Arbuthnot commanded the naval component of the expedition. Despite their conflicting personalities, Clinton and Arbuthnot managed to execute a superb example of a joint operation, landing Clinton's army on the Sea Islands south of Charleston and then supporting the land forces as they advanced toward the city. Joined by almost two thousand men from the Savannah garrison, Clinton bottled up Maj. Gen. Benjamin Lincoln's force of almost six thousand men in Charleston, which included the regular regiments from South Carolina, North Carolina, and Virginia—almost a third of the Continental Army. During a siege that lasted from 30 March until 12 May, few casualties were suffered by either side, and Clinton implored Lincoln to surrender.[22] After being rebuffed several times, Clinton was prepared to lay waste to the city, but Lincoln, persuaded by the Charleston political leaders, agreed to surrender. A frustrated Clinton refused to allow Lincoln's

---

[19] Brown, "James Simpson," p. 513; Borick, *Gallant Defense*, pp. 16–24.

[20] Brown, "James Simpson," p. 519.

[21] Wilson, *Southern Strategy*, p. 198; Borick, *Gallant Defense*, p. 23.

[22] Total casualties were 99 killed and 217 wounded among the British Army and naval forces. The Americans lost 89 killed and 138 wounded, mostly from the Continental forces. Borick, *Gallant Defense*, p. 222.

army the normal honors of war and humiliated them upon their surrender.[23] Though Lincoln and many of his senior officers were paroled and eventually exchanged, many of the rank and file of the Continentals were forced to endure the filthy prison ships in Charleston harbor for almost a year.[24] The largest city in the South, its center of political power and economic trade, had fallen to a completely successful conventional military campaign.

Phase IV: Postcombat Operations

The British victory on 12 May sent a shock wave through the southern states that stunned supporters of the Whig cause and electrified those Loyalists who had suppressed their beliefs since 1776. Clinton and his commanders were almost euphoric in their descriptions of future operations in South Carolina and Georgia. During this phase of operations, they had to accomplish several tasks. First, they had to extend the presence of British troops from the coast into the interior as far as the backcountry to physically demonstrate that Crown authority had returned. Next, as determined by their assumptions and limited force structure, they would recruit and organize Loyalist militia units so that Georgians and South Carolinians could assist with maintaining order in the region. Perhaps the most critical task was to determine how to deal with the former Whigs—an issue that raised questions about how many would return their allegiance to Britain, in what capacity they could serve, and how many would continue to resist, either passively or violently, the British attempt at reconstruction.[25]

To accomplish the first task, Clinton, assisted by his deputy, Lt. Gen. Charles, Earl Cornwallis, sent regular and provincial forces into the interior regions of South Carolina, fanning out from the coast and along the Savannah River into Georgia. While moving forward, the commanders of the British units had orders to destroy any remnants of Rebel forces and encourage the Loyalists to take control of local regions. During these early deployments, the inhabitants and the British military forces had their first personal interaction with each other. In some cases, the regular troops, who generally lived at poverty level, pillaged and looted farms without concern for the owner's status as a Whig or Loyalist. When Whigs were victims, it was deemed a fortune of war, but acts against Loyalists could result in the making of new enemies. Loyalist columns also used this opportunity to seek retribution to settle old scores. The

---

[23] This action was viewed as a personal insult to Benjamin Lincoln, but it would be repaid at Yorktown on 19 October 1781, when Cornwallis' army surrendered under similar terms to Washington's deputy commander, Benjamin Lincoln.

[24] Borick, *Gallant Defense*, p. 223.

[25] A good overview of these problems is described in Louis D. F. Frasche, "Problems of Command: Cornwallis, Partisans and Militia, 1780," *Military Review* (April 1977): 60–74.

temporary void of power between the Whig regime and restored Crown rule also contributed to a general lawlessness and rise of criminal activity.[26]

One incident made a lasting impression on all inhabitants. On 29 May 1780, a British column led by Lt. Col. Banastre Tarleton and a small Continental force led by Col. Abraham Buford fought an action in a region known as the Waxhaws near the North Carolina border. After the fight ended, 113 of Buford's men lay dead, with 150 wounded and 53 captured. The British lost only 17 casualties. The Whigs alleged that Tarleton had ordered surrendering American prisoners killed and promoted a perception that the British did not adhere to the laws of war. Tarleton soon became known as "Bloody Ban" as Whig sympathizers sought to use incidents such as Waxhaws to sway opinion away from supporting the Crown.[27]

As Colonel Brown led his veteran King's Carolina Rangers north from Savannah to occupy Augusta, he did not meet any resistance.[28] On 18 June, Brown reported to Cornwallis that he had taken Augusta and had initiated efforts to restore royal authority to the backcountry. On the same day, Lt. Col. Nisbet Balfour moved to possess the post at Ninety-Six, while another force under command of Cornwallis moved to occupy Camden. The Whigs offered no resistance to these operations, and the British believed that they had defeated all organized opposition.[29]

The British secured their gains in the interior by establishing a series of interconnected posts—what we would call forward operating bases, or FOBs, today— along key lines of communication, such as rivers and roads. These posts began at the coast and included Savannah, Charleston, and Georgetown; extended through Camden, Rocky Mount, and Hanging Rock; and along the Savannah River from Ebenezer to Augusta to Ninety-Six. Brown's base at Augusta was critical because it secured the British left flank along the Savannah River, and it served as the gateway to trade and communication with the Cherokee and Creek nations. Additional smaller fortified posts such as Fort Granby, Fort Motte, and Fort Watson linked these bases into a network for logistics and communication throughout South Carolina and Georgia.[30]

---

[26] Pancake, *Destructive War*, pp. 94–95.

[27] Rpt, Tarleton to Cornwallis, 30 May 1780, in Lt. Col. Banastre Tarleton, *A History of the Campaigns of 1780 and 1781 in the Southern Provinces of North America* (1787; repr. North Stratford, N.H.: Ayer Company Publishers, 2001), p. 84. The most recent study of the Waxhaws battle is by Thomas A. Rider, "Massacre or Myth: No Quarter at the Waxhaws, 29 May 1780" (Master's thesis, University of North Carolina at Chapel Hill, 2002). In addition, see James Piecuch, "Massacre or Myth? Banastre Tarleton at the Waxhaws, May 29, 1780," *Southern Campaigns of the American Revolution* 1, no. 2 (October 2004): 3–10, available at www.southerncampaign.org; Wilson, *Southern Strategy*, pp. 242–61.

[28] Brown to Cornwallis, 18 Jun 1780, 30/11/2, Cornwallis Papers, pp. 166–68, British Public Records Office (BPRO).

[29] Henry Lee, *Memoirs of the War in the Southern Department of the United States* (1869; repr., New York: Arno Press, 1969), pp. 163–64.

[30] Tarleton provides a detailed description of the bases and units assigned to garrison and patrol the region. Tarleton, *History of the Campaigns*, pp. 86–88.

These forward bases served several purposes. They could garrison troops sent out to patrol the immediate area and demonstrate the presence of royal authority. They allowed each base to support others, as they were usually within a day or two march or travel by water, and they provided a safe haven where wounded could convalesce, food could be gathered from the countryside and stored, jails that could hold fugitive Rebels, and courts that could try them. The forward bases served as centers for communication of policy and a place of refuge for those harassed where British power was absent.

Collapse of Whig Resistance in the Backcountry

Many Whig militia leaders were convinced that the rebellion was over. On 28 June, Brown and other Loyalist officers received the surrenders of various Whig units. Loyalist William Manson accepted the surrender of Georgia militia Col. John Dooley's command of about 400 men, along with over 210 stands of arms.[31] Brown accepted the surrender of Cols. Benjamin Garden's and Robert Middleton's South Carolina regiments and ordered them to return to their homes without any penalties as prisoners on parole.[32] Prominent regional leaders, such as Andrew Pickens from the Ninety-Six district, who had openly resisted British and Loyalist activity in the preceding years, accepted parole and returned to their plantations with their wills broken. However, a few others, such as Dooley's subordinate, militia Lt. Col. Elijah Clarke of Wilkes County, Georgia, and Maj. James McCall from South Carolina, determined to remain in the field and resist all British efforts to restore royal order throughout the region.[33]

---

[31] Robert S. Davis Jr., "The Last Colonial Enthusiast: Captain William Manson in Revolutionary Georgia," *Atlanta Historical Journal* 28 (Spring 1984): 23–38. For John Dooley, see Robert Scott Davis, "Colonel Dooley's Campaign of 1779," *Huntington Library Quarterly* 46 (Winter 1984): 65–71, and Davis, "A Frontier for Pioneer Revolutionaries: John Dooley and the Beginnings of Popular Democracy in Original Wilkes County," *Georgia Historical Quarterly* 90 (Fall 2006): 315–49.

[32] Edward J. Cashin and Heard Robertson. *Augusta and the American Revolution*. (Darien, Ga.: Ashantilly Press, 1975), p. 42.

[33] There is not a good biography of Elijah Clarke or James McCall. The work by Louise Frederick Hayes, *Hero of the Hornet's Nest: A Biography of Elijah Clarke* (New York: 1946), is more akin to historical fiction than scholarly research. For a reliable summary of Clarke's life, see Robert Scott Davis, "Elijah Clarke: Georgia's Partisan Titan," *Southern Campaigns of the American Revolution* 4, nos. 1–3 (January–March 2007): 37–39, available at www.southerncampaign.org. Even less is available on James McCall, though much may be found in his son's account of Georgia history. See Hugh McCall, *The History of Georgia* (Atlanta: Cherokee Publishing Co., 1909), as well as recent studies by Sam Fore, "Presentation on Lieutenant Colonel James McCall Given at Musgrove's Mill State Historic Site, Clinton, SC, on December 17, 2005," *Southern Campaigns of the American Revolution* 3, no. 1 (January 2006): 11–13, and Daniel Murphy and Ron Crawley, "The Real Life Exploits of an Unknown Patriot: Lt. Col. James McCall," *Southern Campaigns of the American Revolution* 3, no. 12 (December 2006): 19–23, available at www.southerncampaign.org.

Clinton's Proclamation—You Are with Us or Against Us

Clinton sought to shift the burden of reconstruction to the inhabitants as quickly as possible, stating, "The helping hand of every man is wanted to re-establish peace and good government."[34] However, before the British could put a "southern face" on the situation, they had to implement safeguards to ensure that former Whig regime members were eligible to resume important civic responsibilities. This process of "de-Whigification" proved critical to subsequent events. The British insisted that all subjects take public oaths of loyalty before they could be employed in any official political or military capacity.[35] They arrested those Whigs identified as unredeemable either ideologically or due to specific acts they had committed and held them in Charleston or, in some cases, removed them from the region and sent them to prison in St. Augustine, Florida.[36]

Though thousands of South Carolinians and Georgians agreed to take the oath, open demonstration of loyalty often depended on the assured presence of regular British troops.[37] Many Loyalists chose to withhold their full support until they were sure the army had eliminated armed remnants of the Whig regime.[38] Clinton, however, was convinced that calm would be quickly restored and stated, "From every Information I receive, I have the strongest reason to believe that the general Disposition of the People to be not only friendly to the Government, but forward to take up arms in its support."[39]

On 1 June 1780, Clinton issued what amounted to a full pardon for most treasonable offenses, except for murder, committed by any Whigs who surrendered and accepted parole.[40] This policy meant that most former Whigs could return to their homes and, as long as they did not take up arms in rebellion, that they would be accepted back into British society. Unfortunately, this policy aggravated the Loyalists, who had suffered years of abuse and now found their former oppressors restored to full citizenship, with no obligation to demonstrate loyalty through military service. By merely taking an oath, former

---

[34] Handbill issued after the surrender of Charleston, Tarleton, *History of the Campaigns*, pp. 68–70.

[35] Prominent Whigs such as Charles Pinckney, former president of the Continental Congress; Henry Middleton; and one of Lincoln's cavalry commanders, Daniel Horry, took the oath of allegiance. Borick, *Gallant Defense*, p. 232.

[36] A list of thirty of the most wanted inhabitants of Charleston can be found in Tarleton, *History of the Campaigns*, p. 156, and Note B, p. 186.

[37] Colin Campbell, ed., *Journal of an Expedition Against the Rebels of Georgia in North America Under Orders of Archibald Campbell Esquire, Lieut. Col. of His Majesty's 71st Regiment, 1778* (Darien, Ga.: Ashantilly Press, 1981), p. viii.

[38] Coleman, *Georgia in the Revolution*, p. 122.

[39] Clinton to Cornwallis, 29 May 1780, cited in Borick, *Gallant Defense*, p. 234.

[40] Proclamation, issued by Clinton and Arbuthnot , 1 Jun 1780, in Tarleton, *History of the Campaigns*, pp. 75–76.

Rebels escaped punishment and reverted to the same status as loyal subjects who had suffered for years because of their stalwart allegiance. This outcry from the Loyalists induced Clinton to change his parole policy.

On 3 June 1780, Clinton issued a revised policy, effective 20 June, that stated, "It is . . . proper that all persons should take an active part in settling and securing His Majesty's government and delivering the Country from that anarchy which [has] prevailed. . . . All persons . . . who shall afterwards neglect to return to their allegiance and to His Majesty's government will be considered as Enemies and Rebels to the same and treated accordingly."[41] This released the former Whigs from their paroles on 20 June and, after that date, imposed on them as citizens a duty to actively fight against their former comrades, the "dead-enders" like Clarke and McCall who did not accept the parole. The proclamation meant that former Whigs could not be neutral, no matter how much they wanted to be. They had to be either for or against the British Crown. After backing the former Whigs into a corner where they now had to make a choice, Clinton threw that policy grenade into the lap of Cornwallis and departed Charleston for New York two days later on 5 June. Cornwallis would have to deal with the consequences of this policy.

Cornwallis Assumes Control of Operation Southern Campaign III

Upon Clinton's departure, the responsibility for the campaign and reconstruction of the liberated colonies rested upon Cornwallis, who had about 6,369 regular and provincial troops.[42] He indicated his view of occupation by modifying Clinton's proclamations to ensure greater protection for those who had remained loyal to the Crown and to provide more severe punishment to those who had chosen rebellion. During June, Cornwallis issued an order to the commanders of British forward operating bases to ensure that they understood in no uncertain terms his policy toward those who had second thoughts about their allegiance. Cornwallis wrote,

> I have ordered in the most positive manner, that every militia man who has borne arms with us and afterward joined the enemy, shall be immediately hanged. I desire you will take the most rigorous measures to punish the rebels in the district in which you command, and that you obey in the strictest manner, the directions I have given in this letter.[43]

---

[41] Proclamation, issued by Clinton, 3 Jun 1780, in Tarleton, *History of the Campaigns*, pp. 73–74.

[42] Borick, *Gallant Defense*, p. 236. Cornwallis had 6 British regiments, 3 Hessian regiments, and 6 provincial regiments of Loyalists. Mackesy states that in July, the disposition of British forces in South Carolina and Georgia was 6,129 fit for duty out of 8,439 total effectives. Mackesy, *War for America*, p. 346.

[43] Cited in McCall, *History of Georgia*, p. 481.

The organization of Loyalist internal security forces was most significant in the Ninety-Six district, one of the most populous and contested in the region. Colonel Balfour, assisted by Maj. Patrick Ferguson, whom Clinton had appointed inspector general of militia on 22 May, formed seven battalions of Loyalist militia with about four thousand men.[44] Ferguson served as the main adviser or training instructor, charged with taking undisciplined backcountry Loyalists and turning them into a fighting force capable of opposing their Whig counterparts. His major complaint about his new recruits was their ill discipline—many of them got homesick and simply left the training camps. Also, in the cliquish circles of the British Army, Ferguson was considered a protégé of Clinton, which made for an awkward relationship with Cornwallis and his supporters. Balfour, a Cornwallis man, said of Ferguson, "His ideas are so wild and sanguine . . . it would be dangerous to trust him with the conduct of any plan." In spite of these challenges, Ferguson approached his mission with diligence.[45]

The British formed other battalions of Loyalist militia along the extensive Broad River to Cheraw line, but many of them proved weak or not fully committed to taking on the responsibility for security. In some cases, they simply could not be trusted. One incident involved a whole regiment of Loyalists who defected to the Whig insurgency. The militia from the districts between the Enoree and Tyger rivers had previously served under Whig commander Col. Andrew Neel, who fled South Carolina when Charleston surrendered. A new Loyalist regiment was organized by Col. Mathew Floyd, who accepted former Whig officer John Lisle as second in command after he swore the oath of allegiance. However, as soon as the British completed issuing arms and ammunition to the regiment, Lisle led most of the men to join Whig Col. Thomas Sumter's command near the Catawba.[46] When Floyd later captured two deserters, he had them hung in accordance with Cornwallis' order, which then prompted the remainder of the regiment to desert and join Sumter as well.[47]

Those closer to the populace did not share the optimism of the senior commanders regarding the expected nature of Phase IV operations. Balfour reflected his frustration with the occupation when he wrote,

---

[44] Lt. Col. H. L. Landers, *The Battle of Camden, South Carolina, August 16, 1780* (Washington, D.C.: Government Printing Office, 1929), pp. 31–33.

[45] Ferguson appeared to face many of the same issues that U.S. military advisers are facing in Iraq today: halfhearted recruits who often do not show up for training or simply go home. A very good analysis of Ferguson's challenges as an adviser can be found in John Buchanan, *The Road to Guilford Courthouse: The American Revolution in the Carolinas* (New York: John Wiley & Sons, 1997), pp. 202–04.

[46] Tarleton, *History of the Campaigns*, p. 93.

[47] Pancake, *Destructive War*, p. 82; Cornwallis to Clinton, 6 Aug 1780, in Tarleton, *History of the Campaigns*, p. 126.

> Things are by no means, in any sort of settled state, nor are our friends, so numerous as I expected, from Saluda to Savannah river, almost the whole district . . . are disaffected and although at present overawed by the presence of the troops, yet are ready to rise on the smallest change—as to their disarming it is a joke they have given in only useless arms and keep their good ones.[48]

A Plea for Fortifications at Augusta

Colonel Brown probably held a more realistic view than Cornwallis of the challenges facing the British occupation during the summer of 1780. Fortifications were foremost in Brown's mind as he sought to ensure Augusta's defense and to protect supplies from any potential Whig raid, which he believed was almost certain given the volatility of the region.[49] Brown requested funding and materials from Cornwallis in order to build a suitable fortification. He sought support from Balfour at Ninety-Six, who emphasized Augusta's strategic importance in a letter to Cornwallis, in which he stated,

> As to the post at Augusta . . . it has been and will continue to be the depot for the Indian business, and . . . is a support to this post, and here, I am clear, a force ought to be kept. . . . I conceive a small work will be necessary, as it is so straggling a village and as there are guns and necessarys on the spot. I should think a work for two hundred men perfectly sufficient with Barracks, and they have six four-pounders on the spot.[50]

In spite of these appeals by commanders intimately familiar with the military, political, and social concerns of their areas of occupation, Cornwallis rebuffed their request. In a 3 July reply to Balfour, he specifically forbade the construction of any permanent wood or brick fortification structures at either Ninety-Six or Augusta and instead authorized only earthen fieldworks as a measure of economy.[51] Cornwallis' response reflected his growing irritation at many similar requests he had received from Georgia Royal Governor Sir James Wright, who had pleaded throughout the summer for more troops, supplies, funding, and permanent fortifications to secure Georgia. Cornwallis told Wright, "So long as we are in Possession of the whole Power and Force of South Carolina, the Province of Georgia has the most ample and Satisfactory

---

[48] Quoted in Pancake, *Destructive War*, p. 81.

[49] Many histories of the Revolution wrongly identify Fort Cornwallis as this structure at this time. In reality, there were no forts, just fortified houses, such as Grierson's. Fort Cornwallis did not exist and Fort Augusta was more a memory than a physical structure. Augusta in summer 1780 had no physical defenses. After these events, during fall 1780, Fort Cornwallis would be built.

[50] Balfour to Cornwallis, 24 Jun 1780, 30/11/2, Cornwallis Papers, pp. 191–96, BPRO.

[51] Cornwallis to Balfour, 3 Jul 1780, 30/11/78, Cornwallis Papers, pp. 3–4, BPRO.

Protection by maintaining a Post at Savannah and another at Augusta, nor can I think myself justified in incurring any further expence on the Army Accounts for the Protection of Georgia."[52] Cornwallis' assessment perhaps reflected his desire to focus resources on his next objective, North Carolina. He saw what he wanted to see rather than acknowledging a realistic understanding of just what his forces faced in the backcountry. As a result, Brown's garrison at Augusta made do with inadequate fortification in case of attack.

To men like Brown, hope was not a method, as he knew the paroled Whig Rebels held uncertain levels of commitment to the oath of allegiance they had submitted to in June. In Wilkes County alone, he knew of more than five hundred former Whig Rebels who grudgingly had accepted that their cause was lost. These, along with the hard-core Rebels, concerned Brown.

On 1 July, the "de-Whigification" of Georgia began when Governor Wright signed the Disqualifying Act, which named and barred 151 leaders of the rebellion who had held office under the Whig government from any position in the restored royal government. This act also prohibited former officials from owning firearms, and they could be arrested and brought before a magistrate to swear allegiance to Great Britain. Anyone not complying with this act could be fined, imprisoned, or impressed into the Royal Navy.[53] Wilkes County, Georgia, became a tinderbox waiting to ignite at the slightest hint of Whig strength or British weakness.

Emergence of Tribal Leaders

The policies of Clinton, Cornwallis, and Wright angered many Whigs, who believed they had no incentive to remain neutral. Without any organized Continental force to oppose the British, several Whigs emerged who sought to continue the fight as partisans or insurgents. These partisan leaders exhibited an influence grounded in local social affiliations, genealogical relationships, and strong personalities, which gave them an almost tribal leadership role that inspired men to follow them. One observer commented on the nature of the backcountry people to follow either Whig or Loyalist regional leaders in this way, "But remove the personal influence of the few and they are [a] lifeless, inanimate mass, without direction or spirit to employ the means they possess for their own security."[54] The measure of success for these men was their ability to attract others to the ranks and retain them for operations primarily through their ability to persuade others, either through appeal to their patriotism, greed, vanity, or need for survival.[55]

---

[52] Cornwallis to Wright, 18 Jul 1780, in Manuscript, Colonial Records of Georgia, 38, pt. 2, pp. 413–14.

[53] Cashin and Robertson, *Augusta and the American Revolution*, p. 43.

[54] Nathanael Greene quoted in Pancake, *Destructive War*, p. 92.

[55] Pancake, *Destructive* War, p. 93.

One of these men, Francis Marion, had served as a lieutenant colonel in the South Carolina Continental line but had been absent at the capture of Charleston. He gathered as many men as he could into a small detachment that lived in the swamps north of Charleston and operated east of the Wateree-Santee-Catawba river line. Marion assumed the role as the "swamp fox," attacking the British supply lines from Charleston to the backcountry and using his proximity to the main British base to gather intelligence and pass it to the other leaders.

Colonel Sumter had previously resigned from the Continental Army and retired to his plantation near Statesburgh, when a raiding force from Tarleton's command burned his home. Sumter, known as the Gamecock, assumed the most prominent role during the summer of 1780, effectively raising large numbers of Whigs to fight in the insurgency against the British. Often supporting Sumter was one of the most aggressive of these tribal leaders, Elijah Clarke of Georgia, who had never accepted parole and energetically continued resistance in the backcountry of Georgia, South Carolina, and North Carolina. He often teamed with James McCall of South Carolina.

Absent from the fighting during summer 1780 was Andrew Pickens, the South Carolina militia colonel who had accepted the British parole in good faith. However, he would later emerge as the most influential backcountry leader after the British burned his plantation in November 1780. Pickens already enjoyed a reputation as a successful partisan because of campaigns against the Cherokee Indians in 1776 and his victory over the Loyalists at Kettle Creek, Georgia, in 1779.

These Whig partisan leaders, either working alone or in loose collaboration with each other, provided stubborn and violent resistance in a region that had been paralyzed by the British victory. They continually fanned the spirit of revolt in the occupied areas and fought a relentless and savage war against their Loyalist neighbors. War in the backcountry was without quarter, and the intensity of the violence stunned outsiders such as Nathanael Greene, who later said, "The whole country is in danger of being laid waste by the Whigs and Tories who pursue each other with as much relentless fury as beasts of prey."[56]

Insurgents Versus Occupiers—Summer 1780

During July–September 1780, the tinderbox designed by British policies and actions could no longer contain the tension between the Whigs, Loyalists, Indians, and British. A civil war erupted that crossed religious, social, economic, and family boundaries and signaled a new phase for British military operations. Each side watched closely for reaction to events, for success or defeat would impact the psychology of participants and either encourage or

[56] Louis D. F. Frasche, "Problems of Command: Cornwallis, Partisans and Militia, 1780," *Military Review* (April 1977): 60–74.

depress support for either faction. Hearts and minds were the goals of each side during that time. Space and time do not allow for a detailed study of important battles; the following is an abbreviated description of those events.[57]

On 12 July at Williamson's Plantation, five hundred Whigs from Sumter's command surrounded and surprised South Carolina Loyalist Capt. Christian Huck, leading a detachment of 115 men from Tarleton's British Legion. The Loyalists were routed and Huck was killed, along with thirty-five of his men and over fifty wounded. The same day, fifty miles west near Cedar Springs, South Carolina, Whig Col. John Thomas, commanding the Spartan regiment, received warning of an impending attack from a woman who had learned of it while visiting Ninety-Six. When the detachment of Ferguson's men attacked Thomas, they ran into a prepared ambush and lost about thirty killed. The fighting continued the next day near Gowen's Old Fort, South Carolina, where Col. John Jones, leading Georgia Whigs to join a Whig force in North Carolina, attacked the remaining Loyalists retreating from Cedar Springs. Without central direction, three different Whig columns, each acting independently, engaged and defeated two separate Loyalist forces. In three days of fighting in the region just south of the North Carolina border, the Whigs had killed over ninety Loyalists while suffering about sixty casualties.[58]

On 25 July 1780, Maj. Gen. Horatio Gates assumed command of the Southern Department in North Carolina and began to organize Continental and militia forces to prove that the Continental Congress had not yet written off the South. Knowing that help was on the way, on 30 July Sumter maintained pressure on British bases by attacking Rocky Mount, held by Loyalist Lt. Col. George Turnbull. Turnbull refused to surrender, and Sumter attacked the fortified position three times but was repulsed with loss of about fifteen men. That same day, Lt. Col. William Davie and his North Carolina Whigs ambushed several companies of North Carolina Royalists on their movement to the British base near Hanging Rock. They killed or wounded most of the Loyalists, and Davie captured the weapons and horses of the enemy. A week later, on 6 August, Sumter joined Davie with over 800 men to attack almost 1,400 men of the British garrison at Hanging Rock. The battle of Hanging Rock lasted over four hours, with heavy casualties suffered on both sides. When many of his men stopped to loot one of the British camps, and, running low on ammunition, Sumter withdrew, leaving the base in British hands.[59] By the first week in August, Cornwallis' occupying army had been attacked nearly a dozen times by insurgents who had killed or wounded nearly five hundred

[57] More detailed descriptions can be found in secondary accounts such as Pancake, *Destructive War*, pp. 91–107; Gordon, *South Carolina*, pp. 89–111; Lumpkin, *Savannah to Yorktown*, pp. 264–68; Buchanan, *Road to Guilford*, pp. 104–208.

[58] Pancake, *Destructive War*, pp. 96–97.

[59] Lumpkin, *Savannah to Yorktown*, pp. 264–65. See Rpt, Cornwallis to Clinton, 6 Aug 1780, in Tarleton, *History of the Campaigns*, pp. 126–28.

of his men.[60] Though the British had much presence, with detachments in dozens of forward bases throughout the region, that presence did not translate into control. In fact, they had moved into areas where Whig support was the strongest.

On 7 August, Lt. Col. John Harris Cruger, Balfour's replacement at Ninety-Six, reported intelligence to Cornwallis that Elijah Clarke was raising a force of several hundred men in Wilkes County, Georgia. Cruger requested that Cornwallis send additional troops to Augusta to augment Brown's command.[61] However, Cornwallis had other matters to contend with, as Gates made his appearance in South Carolina with a Continental force of men from Delaware and Maryland. On 16 August, Cornwallis encountered Gates and his army just north of Camden on the old Waxhaws road. There, in a decisive battle, Cornwallis defeated Gates' army, which suffered a loss of almost seven hundred men. To many British, Loyalists, and Whigs, the British victory at Camden appeared to have destroyed another American army and the last gasp of Rebel resistance.

Following the disaster at Camden, Sumter pulled back his command to regroup near Fishing Creek. Tarleton had picked up Sumter's trail and drove his dragoons forward. On 18 August, Tarleton caught Sumter and eight hundred of his men by complete surprise as they either rested or bathed in the river. Sumter escaped, but he lost about 150 men killed and 300 captured.[62] That same day, Musgrove's Mill on the Enoree River witnessed one of the most violent skirmishes of the summer. There, Col. Isaac Shelby, with a detachment of frontier riflemen, teamed with Elijah Clarke of Georgia and James Williams of South Carolina to attack a Loyalist unit commanded by Col. Alexander Innes. Employing a ruse to bait the Loyalists into pursuit, Innes' force left the post at Musgrove's Mill and crossed the Enoree River, right into a trap. Innes fell wounded, and his men were badly defeated, with about 150 killed and wounded and 70 captured.

Francis Marion also kept the British busy near the swamps close to Charleston. On 20 August at Nelson's Ferry, Marion attacked a British column escorting several hundred Whig prisoners from Camden to Charleston. Marion captured or killed about two dozen British troops and freed 150 of the Maryland Continentals, many of whom joined him to continue the war against the enemy. A few weeks later, on 4 September, Marion struck again and ambushed a detachment of British near a swamp island called Blue Savannah.

The intensity of the warfare surprised Cornwallis, while the combat performance of the Loyalist militia disappointed him. Within about six weeks there had been over fifteen skirmishes and battles between Loyalists and Whigs in South Carolina, often including fighters from Georgia and North

[60] Pancake, *Destructive War*, p. 98.

[61] Cruger to Cornwallis, 7 Aug 1780, 30/11/63, Cornwallis Papers, p. 22, BPRO.

[62] Gordon, *South Carolina*, pp. 94–95.

Carolina. This convinced Cornwallis that the backcountry presented a single operational problem and that the occupation of South Carolina depended upon subduing the North Carolina piedmont as well as stopping the further incursion by insurgents from the north. However, the British effort to support the Loyalists to take control had stretched forces so thin that the Whigs could strike at will anywhere along the chain of forward operating bases and the lines of communication to Charleston. Cornwallis was convinced that the best defense of South Carolina was to carry out offensive operations into North Carolina.

Clarke Prepares to Attack Forward Operating Base Augusta

Colonel Clarke had spent July and August moving through the upper backcountry of Georgia, South Carolina, and North Carolina to participate in any opportunity to fight the British forces. The action at Musgrove's Mill on 18 August 1780 cemented his reputation as a hard, courageous fighter, and Clarke hoped to raise a force of over one thousand men to strike at Augusta and Ninety-Six. He asked his companion in these efforts, Colonel McCall, to recruit among the South Carolina men and bring them to a rendezvous forty miles northwest of Augusta in early September.[63] Recruiting insurgents proved a problem for both Clarke and McCall—the impact of Gates' defeat at Camden once again turned events in British favor. McCall appealed to Pickens to support the continued resistance effort of Clarke; however, Pickens rebuffed him with the argument that the paroles they had accepted were binding unless a violation occurred to justify breaking those bonds of honor. As a result, McCall could only persuade about eighty men to join in operations with Clarke.[64]

Clarke, however, was more successful, mainly because his pleas for men were accompanied by threats. Joshua Burnett, one of those who "volunteered" to join Clarke recalled that "[Clarke] sent word to those who had so surrendered, that if they did not meet him at a certain noted Spring in a wilderness, . . . he would put every one of them to death."[65] These recruiting incentives resulted in about 350 men joining Clarke along with the 80 that McCall was bringing for a total of about 430 men.

Clarke decided to attack Augusta to demonstrate that the rebellion in Georgia was not defeated. He also hoped to seize presents and supplies for the Indians stored at Augusta. Finally, many of his men would welcome an opportunity to attack many of the Cherokee and Creek Indians moving along the trails to Augusta.[66]

---

[63] McCall, *History of Georgia*, p. 482.

[64] Ibid.

[65] Joshua Burnett, "The Pension Claim of Joshua Burnett," ed. Edward J. Cashin, *Richmond County History* 10 (Winter 1978): 16.

[66] McCall, *History of Georgia*, p. 483.

While Clarke gathered his partisans, about 250 Creek Indians, led by Little Prince of the Tuckabatchees, answered Brown's call to join forces with the British at Augusta.[67] There, Brown's provincial troops consisted of his battalion of the King's Carolina Rangers, which numbered about 250 men.[68] He stationed one company of rangers at the Mackay trading post, a white stone structure, where they guarded the Indian presents and supplies. He placed his command post and the other ranger companies about a mile and a half east at Loyalist James Grierson's fortified house and St. Paul's church. In addition, Brown had a small detachment of about twenty-seven men from Lt. Col. Isaac Allen's 3d Battalion, New Jersey Volunteers, who were recovering from wounds received at Musgrove's Mill.[69] Brown also had at least two brass artillery pieces, probably three-pounders. Along with the Indians, Brown commanded about five hundred effective soldiers scattered about the Augusta area.

Brown Versus Clarke—Siege of FOB Augusta (14–18 September 1780)

On 14 September 1780, Clarke approached the unsuspecting Loyalists and Indians by dividing his command into three elements to attack from different directions.[70] By using this tactic, Clarke hoped to surprise the superior force, seize key supplies, and kill or capture as many Indians and Loyalists as possible. Early in the morning, Maj. Samuel Taylor began the attack and surprised the Creeks in their camp just outside of the town. It did not take long for Brown to be alerted, and he immediately dispatched his rangers, along with two small pieces of artillery, toward the direction of the fighting. While Brown moved to support the Indians, Clarke and McCall entered Augusta and released more than seventy Rebels who had been in jail; seized Indian presents valued at £4,000; and liberated much of the arms and ammunition that had been turned over previously by surrendering Whig forces.[71] Clarke then moved toward the

---

[67] Cashin, *The King's Ranger*, p. 114.

[68] This is an approximation based on the strength figures for the King's Carolina Rangers from muster rolls dated 29 November 1779 at Savannah and cited in Cashin, *The King's Ranger*, app. Muster Rolls, King's Rangers, pp. 249–93. The rolls from Savannah in late 1779 provide the closest known strength for the assigned companies of Thomas Brown (64); Capt. Andrew Johnston (62); Capt. Joseph Smith (61); Capt. Alexander Wylly (42); and Capt. Samuel Rowarth (41). The actual number fit to fight is unknown, but both Tarleton and Henry Lee state in their memoirs that Brown had about 150 provincials in the fight. See Tarleton, *History of the Campaigns*, p. 161, and Lee, *War in the Southern Department,* p. 199.

[69] Cashin and Robertson, *Augusta and the American Revolution*, n163, pp. 95–96. See also New Jersey Volunteer Return, 30/11/103, Cornwallis Papers, folio 4, BPRO. Available at the On-Line Institute for Advanced Loyalist Studies, www.royalprovincial.com/military/rhist/njv/njvretn3.htm.

[70] McCall, *History of Georgia*, p. 483; Cashin, *The King's Ranger*, p. 115.

[71] Louis Frederick Hayes, *Hero of the Hornet's Nest: A Biography of Elijah Clarke* (New York, 1946), p. 100.

Mackay house, where his men engaged Capt. Andrew Johnston's company and gained possession of the house and all of the supplies.[72]

Meanwhile, Brown had joined the battle with his rangers and soon found himself caught fighting Taylor's forces bearing down the Creek path from the west and Clarke's forces behind him to the east.[73] The Loyalist attack inflicted several casualties on the Whigs, who were beaten back and driven from the house and surrounding outbuildings. In the confusion of the fight, however, some of Clarke's men had used the cover of brush to move around a flank and capture one of the Loyalists' cannons.[74] Clarke managed to direct fire on Brown's position at the Mackay house until early afternoon, then many of his men quietly departed the battle to seek plunder from Augusta.

Brown used this pause in action to improve his defensive position. Since the house was too small to hold the rangers and the Creeks, he directed the Indians to dig earthworks around the perimeter to improve their position. Brown ordered Loyalist Sir Patrick Houstoun (brother of John Houstoun, Whig governor of Georgia) to Ninety-Six with a message for Cruger to send assistance to help drive off the Rebels.[75] By nightfall, Brown and his men were well-established in a good defensive position and were prepared to meet a renewed Rebel attack. Early the next morning, about fifty Cherokees crossed the Savannah River and joined Brown's forces at the trading post. Brown continued to send written updates to Cruger while his men continued to improve their defensive positions.[76] In his message to Cruger, Brown stated, "I shall defend my post to the last extremity."[77]

About noon, the Rebels opened fire with artillery, which did some damage to the Mackay house. Clarke also directed rifle and small arms fire during a fusillade in which Brown was hit in both thighs by a rifle bullet. Though he was knocked down and in great pain, Brown continued to direct the defense.[78] By early evening on 15 September, dead and wounded men covered the area surrounding the British position. At Ninety-Six, Cruger received the message about the attack and sent a report to Cornwallis about the emergency.[79] At Augusta, Clarke sent Brown a message under a white flag demanding that he surrender, but Brown rejected the demand, promising that Clarke's actions would bring retribution to him, his followers, and their families. With that final

[72] Cashin, *The King's Ranger*, p. 115.

[73] This location today is approximately where Eve and Broad Streets meet. The Ezekiel Harris house sits on the northeast slope of what remains of Garden Hill.

[74] Cashin, *The King's Ranger*, p. 116.

[75] Lt Col Thomas Brown to Lt. Col. John Harris Cruger, 15 Sep 1780, 30/11/64, Cornwallis Papers, pp. 65–66, BPRO.

[76] This number reflects all potential Crown strength in Augusta at this time and includes those fit for duty, wounded, missing, absent (such as messengers), and dead.

[77] Brown to Cruger, 15 Sep 1780, 30/11/64, Cornwallis Papers, pp. 65–66, BPRO.

[78] Cashin, *King's Ranger*, p. 116.

[79] Cruger to Cornwallis, 15–16 Sep 1780, 30/11/64, Cornwallis Papers, pp. 67–68, BPRO.

rejection, the Whigs opened up with a burst of fire upon the Loyalist position and continued the firing throughout the night.

The siege of Augusta continued through 17 September. Brown conducted his defense under extremely aggravating conditions due to the heat and lack of food and water. Though only a few hundred yards from the Savannah River, the British were cut off by the Rebels from all sources of water. In a decision that reflected imagination, resolve, and desperation, Brown ordered his men to preserve their urine in some stoneware. When the urine became cold it was issued out to the men, with Brown himself taking the first drink. For food, all the Loyalists had to eat were raw pumpkins. Added to these discomforts was the stench of dead men and horses and the wailing cries of the wounded calling for water and aid.[80] During all this time, Brown, whose wounds grew more aggravating, continued, "at the head of his small gallant band, directing his defence, and animating his troops by presence and example."[81]

Meanwhile, at Ninety-Six, Cruger had departed with a relief force during the morning of 16 September. Marching toward Augusta was his first battalion of Delancy's New York provincials; a detachment from Colonel Allen's 3d Battalion, New Jersey Volunteers; and Colonel Innes' South Carolina Royalists, a force of about three hundred men.[82] It would take Cruger almost forty-eight hours to reach Augusta and assist Brown. During this time, Clarke's already small force may have been reduced to about two hundred men who still focused on the mission, as many others sought plunder in the town.

At about 0800 hours on 18 September, Cruger's column appeared within sight of Augusta. The arrival of three hundred fresh Loyalists was enough to induce many Rebels to flee from the battlefield. Brown ordered his troops to sally out from their works to capture any stragglers. By that time, Clarke decided he had accomplished all he could and ordered his men to break off the engagement and rendezvous at Dennis Mill on the Little River.[83] Clarke and his men had to run for their lives and, as Brown had promised, the consequence of his insurrection would affect the homes and families of those who had chosen to participate in the attack upon the British base.

Due to their exhausted physical condition after four days of siege, Brown's rangers were unable to pursue the Rebels far, but they did manage to recover the artillery and capture several wounded Rebels. The Creeks and Cherokees moved quickly to capture and kill as many of the Rebel stragglers as they could, along with seizing horses and weapons. In the end, Brown and Cruger's combined

---

[80] Jones, *History of Georgia*, p. 457.

[81] An unusual admirer of Brown was Lt. Col. Henry "Lighthorse Harry" Lee, who often wrote of his enemy's tenacity, courage, and determination in glowing terms. Lee, *War in the Southern Department,* pp. 199–200.

[82] Cruger to Cornwallis, 16 Sep 1780, 30/11/64, Cornwallis Papers, BPRO.

[83] Cashin, *King's Ranger*, p. 118.; Hayes, *Hero of Hornet's Nest*, p. 101.

forces had killed or wounded about sixty of Clarke's men.[84] The Loyalists lost an unknown number killed and the Indians lost about seventy killed in the action.[85] Cruger reported to Cornwallis on 19 September, "I got here yesterday morning. . . . I am now sending out patrols of horse to pick up the traitorous rebels of the neighborhood, who I purpose to send to Charles Town."[86]

Hangings at the "White House"

Perhaps the most well-known incident related to this battle concerned the fate of the Whig prisoners. In accordance with Cornwallis' policy about those who broke their parole and took up arms, the Loyalist commanders were compelled to take action toward thirteen of the captured men. They hanged Captain Ashby from McCall's South Carolina militia and twelve others from an outside staircase of the Mackay house for having participated in the recent battle.[87]

Whig histories have turned this event into a "Waxhaws" of sorts for Brown, and it has tainted, deservedly or not, his reputation. South Carolina Governor John Rutledge even used the "Thomas Brown defense" to justify executing Loyalist prisoners following the Battle of King's Mountain.[88] Nineteenth-century historians such as Charles Jones described how Brown's injuries dictated that the Rebels be "hung upon the staircase of the White House, where Brown was lying wounded, that he might enjoy the demoniacal pleasure of gloating over their expiring agonies."[89] Hugh McCall, son of James McCall, described Brown as having "the satisfaction of seeing the victims of his vengeance expire."[90] However, these descriptions appear contradictory to Brown's character and career. In fact, a strong case can be made that Cruger ordered enforcement of law and supervised the hangings, as Brown would have been incapacitated, having suffered from the stress of command during four days of siege, painful wounds in both legs, and having subsisted on a diet of

[84] Known Whig dead included Capt. Charles Jourdine, Capt. William Martin, Absalom Horn, William Luckie, and Major Carter.

[85] Among the known dead of the Loyalists were Capt. Andrew Johnston and Ensign Silcox of the rangers.

[86] Cruger to Cornwallis, 19 Sep 1780, 30/11/64, Cornwallis Papers, pp. 104–05, BPRO.

[87] The list of names and their spellings varies in the sources. I have chosen to rely on the list supplied by Cashin in footnote 48, p. 315, of *King's Ranger*. McCall cites as his sources British officers who witnessed this event and had "exultingly communicated it" to their friends in Savannah, Charleston, and London. McCall, *History of Georgia*, p. 487.

[88] Rutledge said, "It is said (and I believe it) that of the Prisoners whom Brown took at Augusta, he gave up four to the Indians who killed em, cut off their Heads and kicked their bodies about the Streets and that he (Brown) hung upwards of 30 prisoners" (cited in Cashin, *King's Ranger*, p. 120).

[89] Jones, *History of Georgia*, p. 458.

[90] McCall, *History of Georgia*, p. 486.

pumpkins and urine.[91] If it was in fact Brown, gleeful or not, he ensured the enforcement of Cornwallis' policy.

Loyalist Retribution Against Georgia Whigs

Governor Wright's concern about the state of military security in Georgia was confirmed by the Whig attack upon Augusta, which demonstrated that the "the Spirit and Flame of Rebellion was not over." He again urged construction of proper defensive fortifications at Augusta. In addition, he advised British military leaders that "the most Effectual and Best Method of Crushing the Rebellion in the Back Parts of this Country, is for an Army to march without Loss of time into the Ceded Lands—and to lay Waste and Destroy the whole Territory."[92]

Cruger took command of subsequent operations to hunt down the remnants of Clarke's force and discourage another such insurrection. On 20 September, he received intelligence that Clarke had retreated as far north as the Little River, where he was regrouping for another attack on Augusta after the British moved back to Ninety-Six. Cruger decided to take the fight directly into the backcountry and sent detachments in all directions to mete out frontier justice to the insurgents, their families, and any others who demonstrated sympathy for the Whig cause.

By 23 September, Cruger's force had reached John Dooley's farm about forty-five miles north of Augusta, but, by then, Clarke had already crossed north across the Broad River into South Carolina. Following Wright's advice, the Loyalists under Cruger inflicted a terrible retribution for Clarke's attack. In Wilkes County, the courthouse was burned, frontier forts destroyed, and over one hundred Whig homes were burned, their property plundered, and livestock driven off. The families of the men who had joined Clarke were given a choice of leaving the colony within twenty-four hours or taking an oath and submitting to the royal government.[93] As he pursued the Rebels, Cruger ordered many arrests. Whigs who had been on parole were arrested and sent to Charleston for confinement.[94] By the time Cruger reached the Broad River on 28 September, he could find no trace of the Rebels, who had fled toward the mountains of North Carolina.

After leaving Augusta, Clarke and the remnants of his followers scattered to their homes to gather their families and prepare to leave Georgia for refuge in North Carolina. At an appointed rendezvous, over three hundred men and four hundred women and children met for the arduous journey, carrying only five days of supplies. One historian characterized this event, "Like Moses

---

[91] Cashin makes an important point that officers, such as Henry Lee, expressed admiration for Brown in their memoirs, which perhaps reflects a contemporary view of reality rather than postwar myth (Cashin, *King's Ranger*, p. 120).

[92] Cited in Robertson, *Second British Occupation,* p. 435.

[93] Cashin and Robertson, *Augusta and the American Revolution*, p. 50.

[94] McCall, *History of Georgia*, pp. 488–89.

from Egypt . . . Colonel Clark commenced a march of near two hundred miles, through a mountainous wilderness," to reach the Watauga Valley.[95] Cruger reported Clarke's flight toward North Carolina to Cornwallis, who directed Major Ferguson, with his 1,100-man Loyalist force operating in western South Carolina, to intercept Clarke. Ferguson eventually established a position at King's Mountain to block Clarke and to discourage further rebellion in that region. Clarke, however, escaped, and Ferguson was surprised by a force of three thousand "over-mountain" men, who attacked his position on 7 October in one of the most decisive battles of the war.[96] Later, Cornwallis wrote Clinton about Ferguson's defeat at King's Mountain, stating, "Maj. Ferguson was tempted to stay near the mountains longer than he intended, in hopes of cutting off Col. Clarke on his return from Georgia. He was not aware that the enemy was so near him, and in endeavoring to execute my orders of . . . joining me at Charlottetown, he was attacked by a very superior force & totally defeated at King's Mountain."[97] For Cornwallis and the men of the British forces, the war had entered a new phase that replaced the optimism that had characterized their operations only a few months earlier. King's Mountain signified a turning point in the war in the South and reflected that the assumptions of operating in the southern theater needed to be reassessed by the British leadership.

Significance of May–September 1780

The British experience in attempting to conduct Phase IV operations during the summer of 1780 reflected either complexities and challenges they failed to understand or a hubris that could not conceive of such events. At its heart, the occupation plan may have been sound militarily, but it did not adequately consider how the events of 1776–1780 had fractured beyond repair the relationships between the Loyalists and the Whigs. To mend that rift would have required occupation policies designed to address the grievances of each side so that resentment did not boil into civil war. Clinton's policies did not accomplish that objective and instead may be identified as one of the fundamental causes of the internecine war that began in July 1780. In lieu of policies to mitigate resentment, the presence of larger numbers of British troops was needed if for nothing more than to protect the Loyalists and to convince the Whigs of the futility of further resistance. Instead, stability of local areas was determined by the presence or absence of British troops. Because the limited British forces were spread out into small detachments occupying a vast network of forward operating bases, they were open to insurgent attacks, such as occurred at Augusta. Though the British had much presence throughout the region, they did not have the level of control that encouraged widespread

---

[95] McCall, *History of Georgia*, pp. 490–91.

[96] Cashin and Robertson, *Augusta and the American Revolution*, p. 50.

[97] Cornwallis to Clinton, 3 Dec 1780, 30/11/72, Cornwallis Papers, pp. 57–64, BPRO.

support by those who were at best lukewarm in their support of the Crown. Without that assurance, declaring loyalty exposed one to a potential death sentence at the hands of Whigs.

The battles that summer by Whig insurgents reflected the tribal nature of the region, as men like Marion, Sumter, and Clarke rallied men to their cause to disrupt and discredit British reconstruction efforts. Not helping the British cause was a weak system for recruiting and training Loyalist militia, whose performance may have reflected the lack of adequate British forces to help train and bolster their self-confidence. Men like Brown, Wright, Balfour, and Cruger, who understood the volatility of the region, were refused resources they needed because their views did not fit into preconceived notions for the overall campaign.

The example of Augusta may not seem all that significant in the greater scope of the American Revolution; however, one result was a changed outlook in the backcountry on the part of both Loyalists and Whigs. In one sense, Clarke's attack was a ringing endorsement of the arguments made by Wright, Brown, Balfour, and Cruger. Clarke opened Cornwallis' eyes regarding the need of fortifications at Augusta, something Brown could never accomplish no matter how rational his argument. After Clarke's attack, the gloves came off, and Cruger directed a punitive expedition into the backcountry against persons and property identified with the Rebel cause. The Loyalists were not going to allow an attack like that to happen again against a key British operating base. The action at Augusta also exercised the working relationships of the Loyalist commanders, who saw their roles as mutually supporting reaction forces who would come to each other's aid in checking any Rebel operations.

On the other hand, the Whig cause may have gained momentum in some respects due to the post-battle events. The execution of Rebel prisoners, regardless of the legality of the sentence, served the Whig cause far beyond the vicinity of Augusta. The Whig press and information network spread the news of this event, painting Brown as the devil incarnate, an example of the barbaric British occupation, and justification for retaliation in kind for Loyalist prisoners. Such an event, while demoralizing in one sense, served to harden the resolution of many Whigs, certainly those related to the men and their families who suffered retaliation for the attack upon Augusta. When Clarke and the hundreds of displaced men, women, and children made their way through Georgia, South Carolina, and into North Carolina, their status as refugees served as a further example of the cruelty inflicted by Loyalist punitive actions. All of these aspects may have contributed to making more Rebels in the backcountry rather than convincing people to declare loyalty to the Crown.

Finally, because Clarke and his Georgians fled to North Carolina, Patrick Ferguson was ordered to intercept the Rebels as they retreated to the mountains. However, instead of Clarke, he found an assembly of militia from Western Virginia and North Carolina, who turned their sights on him at a place called

King's Mountain. King's Mountain was the culmination of events of the summer of 1780 that reflected the British failure to understand the complex cultural, political, social, and psychological nature of the enemy they were fighting. The Whigs successfully disrupted British Phase IV efforts to build a secure and stable environment by striking at their weak points, intimidating their neighbors through threats and violence, and portraying the British and Loyalist forces as killers of the innocent. The reasons for their success are reflected in the words of T. E. Lawrence, who later said,

> Rebellion must have an unassailable base, . . . in the minds of the men we converted to our creed. It must have a sophisticated alien enemy, in the form of a disciplined army of occupation too small to fulfill the doctrine of acreage. . . . It must have a friendly population, not actively friendly, but sympathetic to the point of not betraying rebel movements to the enemy. Rebellions can be made by 2 per cent active in a striking force, and 98 per cent passively sympathetic. . . . Granted mobility, security, time, and doctrine, victory will rest with the insurgents.[98]

[98] T. E. Lawrence, "The Evolution of a Revolt," *Army Quarterly and Defense Journal* (October 1920), repr., Leavenworth, Kans.: Combat Studies Institute, 1999.

When Freedom Wore a Red Coat: How Cornwallis' 1781 Virginia Campaign Threatened the Revolution in Virginia

Gregory J. W. Urwin

Nearly every schoolchild in the United States has heard of the siege of Yorktown. It was the decisive battle that all but ended the military phase of the American Revolution and guaranteed the thirteen colonies' independence. Yorktown represents George Washington's finest hour as a general and the crowning achievement of his ragged Continental Army. It was also the event that assured British Lt. Gen. Charles, Earl Cornwallis, an undeserved place on history's list of famous losers, just as it furnished Americans with an exaggerated view of their martial prowess. President Ronald Reagan helped preside over the ceremonies marking the bicentennial of that pivotal event on 19 October 1981. This was Reagan's first extended, open-air appearance since surviving an assassination attempt the previous March, but he rose to the occasion with the uplifting rhetoric that had already become his trademark. A crowd of 60,000 heard the president evoke the exultation felt by Patriots of Washington's day when he called Yorktown "a victory for the right of self-determination. It was and is the affirmation that freedom will eventually triumph over tyranny." Standing beside Reagan behind a massive shield of clear, bulletproof plastic, French President François Mitterrand politely echoed his host's sentiments by proclaiming Yorktown "the first capital of human rights."[1]

Few Americans would quarrel with Reagan's and Mitterrand's words. History, however, is a matter of perceptions, and sometimes those perceptions are too narrow. Such is the case with Yorktown. American scholars are generally so intent on memorializing Washington's brilliant generalship during the Yorktown campaign that they ignore how close Cornwallis came to subduing Virginia.[2] They also fail to see that there was a dark side to Washington's

---

[1] First quote from *Time*, 2 Nov 1981, p. 31. Second quote from *Virginian-Pilot* (Norfolk), 20 Oct 1981. This paper is a slightly expanded and retitled version of the one read at the conference of Army historians held in Arlington, Virginia, in August 2007. Anomalies in spelling, punctuation, and capitalization found in the materials quoted in this article have been retained without comment.

[2] Among the many triumphalist histories of the Yorktown campaign are Henry P. Johnson, *The Yorktown Campaign and the Surrender of Cornwallis, 1781* (New York: Harper & Bros., 1881); Thomas J. Fleming, *Beat the Last Drum: The Siege of Yorktown, 1781* (New York: St.

celebrated triumph. Yorktown meant liberty and independence for the majority of the young republic's white citizens, but it signified something else for the 500,000 blacks who lived in the United States in 1781. For most African Americans, Yorktown meant another eighty years of chattel slavery. And for many of the freedom-loving blacks who cast their lot with the British and joined Cornwallis in the summer of 1781, Yorktown became not merely the graveyard of their hopes, but of their mortal remains.

It seems unfair to say the British lost the Revolutionary War, for they never quite realized what they were up against. To George III and his advisers, the rebellion was a plot hatched by an evil minority, opportunistic demagogues who deluded the riffraff of the thirteen colonies into opposing lawful government. The British sincerely believed that most upstanding Americans remained loyal to their king. All that was required to quell the uprising was a show of force to discredit Rebel leaders and frighten America's masses into resuming their proper allegiance.[3]

Since the British were out to win hearts and minds, they usually did not treat Americans with the same cruelty they reserved for rebels in Catholic Ireland or the Scottish Highlands. Unrestrained barbarism would cost the Crown potential American supporters and even alienate committed Loyalists. As the British were so sure the Revolution had no legitimate appeal, they did not act with the energy or ruthlessness that the situation warranted.[4]

Martin's Press, 1963); Burke Davis, *The Campaign That Won America: The Story of Yorktown* (New York: Dial Press, 1970); Joseph P. Cullen, *October 19th, 1781: Victory at Yorktown* (Washington, D.C.: U.S. Department of the Interior, 1976); William H. Hallahan, *The Day the Revolution Ended, 19 October 1781* (New York: Wiley, 2004); Benton Rain Patterson, *Washington and Cornwallis: The Battle for America, 1775–1783* (Lanham, Md.: Taylor Trade Publishing, 2004); Richard M. Ketchum, *Victory at Yorktown: The Campaign That Won the Revolution* (New York: Henry Holt, 2004). A relatively recent drum-and-bugle account of the campaign with more of a British perspective is Brendan Morrissey, *Yorktown 1781: The World Turned Upside Down* (London: Osprey, 1997). The latest and most scholarly treatment of this subject is Jerome A. Greene, *The Guns of Independence: The Siege of Yorktown, 1781* (New York: Savas Beatie, 2005). Unfortunately, Greene is unable to fully free himself of the triumphalist tendencies that influenced the work of his predecessors.

[3] Piers Mackesy, *The War for America, 1775–1783* (Cambridge, Mass.: Harvard University Press, 1964), pp. 12–61; Ira D. Gruber, *The Howe Brothers and the American Revolution* (New York: Atheneum, 1972), pp. 3–88; Paul H. Smith, *Loyalists and Redcoats: A Study in British Revolutionary Policy* (Chapel Hill, N.C.: University of North Carolina Press, 1964), pp. 10–42; Troyer Steele Anderson, *The Command of the Howe Brothers During the American Revolution* (New York: Oxford University Press, 1936), pp. 10–42; John Richard Alden, *General Gage in America* (Baton Rouge, La.: Louisiana State University Press, 1948), pp. 151–91, 205–50, 287–98; Marshall Smelser, *The Winning of Independence* (Chicago: Quadrangle Books, 1972), pp. 58–117.

[4] John Richard Alden, *A History of the American Revolution* (New York: Alfred A. Knopf, 1969), pp. 141–225; Frederick Mackenzie, *Diary of Frederick Mackenzie*, 2 vols. (1930; repr., New York: New York Times and Arno Press, 1968), 2:525–26; John Shy, *A People Numerous and Armed: Reflections on the Military Struggle for American Independence* (New York: Oxford University Press, 1976), pp. 202–22; *Collections of the New-York Historical Society*,

The British set the basic pattern of the War of Independence during the 1776 campaign in New York and New Jersey. Whenever one of the king's generals wished to conquer a colony, he would head for its largest port, defeat whatever American army stood in his way, occupy his objective, establish a network of outlying outposts, and then wait for the Rebel cause to come unglued. That never happened. The beaten Continental forces would simply retire beyond easy reach, recruit themselves up to strength, and then take positions that threatened the enemy's smaller and more isolated outposts with sudden capture. At the same time, inflamed local militia harassed British garrisons and foraging parties, giving the occupiers no rest and depriving them of any sense of security. Forced to concentrate to avoid defeat in detail, the British found themselves confined to a few major towns and living under virtual siege.[5]

With the Rebels controlling most of the countryside, Loyalists found it impossible to rise in decisive numbers. Any Tory who openly declared for the king risked the loss of his property, imprisonment, and possibly death. Rather than chance such perils, many Loyalists adopted a wait-and-see attitude. If the king's regulars were victorious, loyal subjects would lose nothing by their silence while the issue teetered in the balance.[6]

To break the stalemate that came to characterize the American war, royal commanders seized more cities, but that strategy gained them nothing except worthless real estate. When a British army tried to divide the colonies by marching down the Hudson in 1777, it was trapped and forced to surrender at Saratoga. That stunning Rebel victory brought France into the war on the side of the United States, and Spain and the Netherlands soon followed suit. Britain now faced a world war, and it strained its military resources to the limit while endeavoring to safeguard a far-flung empire.[7]

Assured that vast numbers of Loyalists inhabited the South, the British decided to shift their operations to Georgia and the Carolinas. In May 1780, General Sir Henry Clinton, the commander-in-chief of His Majesty's Forces in North America, captured Charleston, South Carolina, and more than six

---

vols. 16–17, *The Stephen Kemble Papers, 1773–1789* (New York: New York Historical Society, 1883–84), 16:97–99, 263–64, 269–70, 287.

[5] David Hackett Fischer, *Washington's Crossing* (New York: Oxford University Press, 2004), pp. 66–360; John W. Jackson, *With the British Army in Philadelphia 1777–1778* (San Rafael, Calif.: Presidio Press, 1979), pp. 1–52, 81–106; Alden, *American Revolution*, pp. 262–327; Gruber, *Howe Brothers*, pp. 389–93, 397–400; Mackenzie, *Diary*, 2:525. See also Wayne Bodle, *The Valley Forge Winter: Civilians and Soldiers in War* (University Park: Pennsylvania State University Press, 2002).

[6] Shy, *People Numerous and Armed*, pp. 211–22; Smith, *Loyalists and Redcoats*, pp. 36–37, 60–77, 79–85, 115–21, 168–74.

[7] Sir Henry Clinton, *The American Rebellion*, ed. William B. Wilcox (New Haven, Conn.: Yale University Press, 1954), p. 86; Alden, *American Revolution*, pp. 284–327, 370–88.

thousand Patriot troops whose commander had opted foolishly to defend the doomed port.[8]

Clinton soon returned to his main base at New York City, leaving Cornwallis and 8,000 regulars to establish British rule in the Carolinas. Cornwallis was a robust forty-one years of age when he assumed this important command. He carried himself with the easy self-assurance that sprang from an aristocratic background and twenty-three years of military experience. The earl had been fighting the American Rebels since 1776, and he was esteemed as one of the king's ablest and most aggressive generals.[9]

At the outset, Cornwallis' mission in the Carolinas seemed easy. The capture of an entire Continental army at Charleston left local Patriots demoralized and vulnerable. As the British advanced inland, the Rebels either fled or switched their allegiance to the Crown. Magnanimous in victory, Cornwallis permitted them to take an oath of loyalty and join his Loyalist militia.[10]

Then in the summer of 1780, the Continental Congress sent a new Rebel army to reclaim South Carolina. Though badly outnumbered, Cornwallis crushed this threat on 16 August 1780 at the Battle of Camden, but his victory had a bittersweet taste. At the approach of the Continental troops, the crypto-Rebels of South Carolina turned on the British. Whole units of "loyal" militia took the arms and equipment they had drawn from royal magazines and defected to the guerrilla bands assembling in the swamps outside Charleston.[11]

---

[8] Ltr, James Simpson to Lord George Germain, 28 Aug 1779, in Alan S. Brown, ed., "James Simpson's Reports on the Carolina Loyalists, 1779–1780," *Journal of Southern History* 21 (November 1955): 517; Charles Stedman, *The History of the Origin, Progress, and Termination of the American War*, 2 vols. (1794; repr., New York: New York Times and Arno Press, 1969), 2:189; Banastre Tarleton, *A History of the Campaigns of 1780 and 1781, in the Southern Provinces of North America* (1787; repr., New York: New York Times and Arno Press, 1968), pp. 13–23; Clinton, *American Rebellion*, pp. 150–74; Alden, *American Revolution*, pp. 408–09, 412–16; Smith, *Loyalists and Redcoats*, pp. 121–28; Mackenzie, *Diary*, 2:447, 525. A reliable biography of the Continental general who lost Charleston is David B. Mattern, *Benjamin Lincoln and the American Revolution* (Columbia: University of South Carolina Press, 1995). The most detailed scholarly account of that campaign is Carl P. Borick, *A Gallant Defense: The Siege of Charleston, 1780* (Columbia: University of South Carolina Press, 2003). Perhaps the most astute evaluation of the ill-founded assumptions that lured the British into the South is David K. Wilson, *The Southern Strategy: Britain's Conquest of South Carolina and Georgia, 1775–1780* (Columbia: University of South Carolina Press, 2005), pp. xi, xv, 59–62, 265–66.

[9] Franklin B. Wickwire and Mary Wickwire, *Cornwallis: The American Adventure* (Boston: Houghton Mifflin, 1970), pp. 8–137; Stedman, *American War*, 2:190–95; Clinton, *American Rebellion*, pp. 174–76, 186.

[10] Stedman, *American War*, 2:190–95; Tarleton, *Campaigns*, pp. 23–32, 71–76, 85–89; Smith, *Loyalists and Redcoats*, pp. 128–36; Wilson, *Southern Strategy*, p. 264.

[11] Ltr, Cornwallis to Clinton, 6 Aug 1780, in *Correspondence of Charles, First Marquis Cornwallis*, ed. Charles Ross, 3 vols. (London: John Murray, 1859), 1:53; Roger Lamb, *An Original and Authentic Journal of Occurrences During the Late American War, From Its Commencement to the Year 1783* (1809; repr., New York: New York Times and Arno Press, 1968), pp. 302, 343, 381; Stedman, *American War*, 2: pp. 198–209; Tarleton, *Campaigns*, pp.

Later in the year, Cornwallis confronted a second American army under Maj. Gen. Nathanael Greene, Washington's most trusted lieutenant. Keeping just beyond reach, the wily Greene goaded Cornwallis into launching a ruinous midwinter pursuit across barren North Carolina.[12] One of the earl's sergeants called the Carolinas "a country thinly inhabited, and abounding with swamps, [that] afforded every advantage to a partizan warfare over a large and regular army."[13] Greene led the earl on a furious chase for nearly two months, finally pausing to fight at Guilford Court House on 15 March 1781. Greene's forces outnumbered the British two to one, but Cornwallis gave battle anyway, and he defeated the Rebels once more. Nevertheless, the outcome of the battle was indecisive, and the cost to the British appalling. Of the 1,900 Redcoats, Hessians, and Loyalists that the earl led into the fray, more than a quarter fell killed or wounded. Another 436 British soldiers suffered bouts of sickness as a result of this strenuous campaign.[14]

Before Cornwallis' ailing army recovered its strength, Greene marched on South Carolina. This time, however, Cornwallis did not join Greene in an exhausting game of cat and mouse. Years of hard campaigning in America had finally shown the earl the flaws in Britain's fundamental strategy. For the rest of that spring and well into the summer—before he received orders to entrench at Yorktown—Cornwallis would experiment with a new approach for subduing the Rebels.[15]

Cornwallis' most significant realization was that most southern Loyalists could not be trusted. "Our experience has shown that their numbers are not so great as has been represented," he wrote ruefully from North Carolina, "and that their friendship was only passive."[16] The Crown's American supporters

---

90–96; Wilson, *Southern Strategy*, pp. 263–65. A fine recent description of Cornwallis' victory is John R. Maass, *Horatio Gates and the Battle of Camden—"That Unhappy Affair," August 16, 1780* (Camden, S.C.: Kershaw County Historical Society, 2001).

[12] A. R. Newsome, ed., "A British Orderly Book, 1780–1781," *North Carolina Historical Review* 9 (1932): 273–98, 366–79; Burke Davis, *The Cowpens-Guilford Courthouse Campaign* (Philadelphia: J. B. Lippincott, 1962), pp. 49–145; Christopher Ward, *The War of the Revolution*, 2 vols., ed. John Richard Alden (New York: Macmillan, 1952), 2:763–83; Tarleton, *Campaigns*, pp. 207–18; Clinton, *American Rebellion*, p. 229. For a recent examination of that campaign, see John Buchanan, *The Road to Guilford Courthouse: The American Revolution in the Carolinas* (New York: John Wiley & Sons, 1997).

[13] Lamb, *Journal*, p. 341.

[14] Henry Lee, *Memoirs of the War in the Southern Department of the United States*, ed. Robert E. Lee (New York: University Publishing, 1870), pp. 276–83. Ltrs, Cornwallis to Rawdon, 17 Mar 1781; Cornwallis to Maj Gen William Phillips, 10 Apr 1781; and Cornwallis to Germain, 18 Apr 1781, last three in Ross, *Correspondence of Cornwallis*, 1:87, 88, 92. Stedman, *American War*, 2: 333–46; Tarleton, *Campaigns*, pp. 271–76; Newsome, "British Orderly Book," pp. 388–92. See also Thomas E. Baker, *Another Such Victory: The Story of the American Defeat at Guilford Courthouse That Helped Win the War for Independence* (New York: Eastern Acorn Press, 1981).

[15] Stedman, *American War*, 2:353; Clinton, *American Rebellion*, p. 284.

[16] During the struggle for the South in 1780 and 1781, the most formidable and steadfast Loyalist units proved to be those that the British imported from the North and not locally raised

talked a good fight, but they usually deserted the royal cause at the first sign of trouble. "The Idea of our Friends rising in any Number & to any Purpose totally failed as I expected," the earl confided to a brother officer, "and here I am getting rid of my Wounded & refitting my Troops at Wilmington."[17] In reference to the handful of southern Tories who attached themselves to his battered army, Cornwallis described them as "so timid and so stupid that I can get no intelligence."[18]

As for the troublesome Greene, the earl had decided that there were less expensive ways to deal with Rebel armies than attacking them directly. Cornwallis would attempt to counter the threat to the Carolinas by striking at the American general's base of supply, the state of Virginia.[19]

Virginia was not only the largest and most populous of the rebellious colonies, but the richest as well. Virginia tobacco was a prime reason why America's staggering economy had not collapsed entirely. With the fall of Charleston, Virginia became the mainstay of the Rebel war effort in the South. It provided the men and materiel Greene needed to keep his army in the field. If Virginia could be knocked out of the war, perhaps the whole Rebel confederation might come tumbling down.[20]

In a letter dated 18 April 1781, Cornwallis expressed his views in these words:

> If therefore it should appear to be the interest of Great Britain to maintain what she already possesses, and to push the war in the Southern provinces, I take the liberty of giving it as my opinion, that a serious attempt upon Virginia would be the most solid plan, because successful operations might not only

---

commands. Wilson, *Southern Strategy*, pp. 89–90, 117, 176, 238, 243, 262–63. Quotes from Ltr, Cornwallis to Germain, 18 Apr 1781, in Ross, *Correspondence of Cornwallis*, 1:90.

[17] Quote from Ltr, Cornwallis to Phillips, 10 Apr 1781, P.R.O. 30/11/85, Papers of Charles Cornwallis, First Marquis and Second Earl, Public Record Office, London (hereafter cited as Cornwallis Papers), microfilm at David Library of the American Revolution, Washington Crossing, Pa. Ltrs, Cornwallis to Clinton, 29 Aug 1780; Rawdon to Clinton, 29 Oct 1780; and Cornwallis to Germain, 18 Apr 1781, the three in Ross, *Correspondence of Cornwallis*, 1:58, 62–63, 89–90. Newsome, "British Orderly Book," p. 372; Tarleton, *Campaigns*, pp. 229–30, 256–57, 312–13.

[18] Ltr, Cornwallis to Lt Col Banastre Tarleton, 18 Dec 1780, in Ross, *Correspondence of Cornwallis*, 1:74.

[19] Ltrs, Cornwallis to Germain, 18 Apr 1781, and Cornwallis to Germain, 23 Apr 1781, both in Ross, *Correspondence of Cornwallis*, 1:90–91, 94–95; Stedman, *American War*, 2:347, 353–54; Lamb, *Journal*, p. 357.

[20] Edmund Randolph, *History of Virginia*, ed. Arthur H. Shaffer (Charlottesville: University Press of Virginia, 1970), pp. 3, 177–78; Ltrs, Richard Henry Lee to Thomas McKean, 25 Aug 1781, in *The Letters of Richard Henry Lee*, ed. James Curtis Ballagh, 2 vols. (1911; repr., New York: Da Capo Press, 1970), 2:247; and Jefferson to Baron Friedrich von Steuben, 13 Jan 1781, in *The Papers of Thomas Jefferson*, ed. Julian P. Boyd, 34 vols. to date (Princeton, N.J.: Princeton University Press, 1950–), 4:352; Lee, *Memoirs*, p. 310.

be attended with important consequences there, but would tend to the security of South Carolina, and ultimately to the submission of North Carolina.[21]

Virginia lay ripe for invasion in 1781. Like other Americans, Virginians were weary after six years of war. Almost all of the Old Dominion's Continental regiments had been captured at Charleston. That left only a few half-trained regulars to defend the state. In addition, large drafts of the Virginia militia had been sent far from home to fight under Greene. Those who survived the arduous campaigns in the Carolinas harbored no desire to face Cornwallis' Redcoats again—a reluctance that they communicated to the militiamen who had stayed behind.[22]

Even nature favored the earl's designs. The most distinctive feature of colonial Virginia's geography was Chesapeake Bay. With its network of great tidal rivers (the James, York, Rappahannock, Potomac, and Susquehanna) and other navigable streams, the Chesapeake served as the highway that brought the first permanent English settlers to North America. It shaped the pattern of Virginia's society and became the key to the colony's prosperity. The Chesapeake also offered an enemy a ready-made invasion route, especially since its twisting, 8,000-mile shoreline was indefensible. As long as the Royal Navy ruled the waves, there was hardly anything of importance in Virginia east of the Blue Ridge Mountains that could not be flattened by British broadsides or menaced by landing parties. Not a town, not a plantation, and not a tobacco warehouse was safe.[23] As Cornwallis astutely observed, "The rivers in Virginia are advantageous to an invading army."[24]

Having taken these facts into account, Lord Cornwallis began his march north toward the Old Dominion on 25 April 1781. By 20 May, he was at

---

[21] Quote from Ltr, Cornwallis to Germain, 18 Apr 1781, in Ross, *Correspondence of Cornwallis*, 1:90, and see also p. 89. See also Ltr, Cornwallis to Germain, 23 Apr 1781, P.R.O. 30/11/5, Cornwallis Papers.

[22] Ltr, Jefferson to Marquis de Lafayette, 23 Apr 1781, in *The Letters of Lafayette and Jefferson*, ed. Gilbert Chinard (Baltimore, Md.: Johns Hopkins Press, 1928), p. 38; Fred Anderson Berg, *Encyclopedia of Continental Army Units—Battalions, Regiments, and Independent Corps* (Harrisburg, Pa.: Stackpole Books, 1972), pp. 124–32. Ltrs, Col William Preston to Jefferson, 10 Apr 1781, in *Calendar of Virginia State Papers*, ed. William P. Palmer, 11 vols. (1880–93; repr., New York: Kraus Reprint Corp., 1968), 2:25, and Richard Henry Lee to Arthur Lee, 4 Jun 1781, in Ballagh, *Letters of Richard Henry Lee*, 2:230. Lee, *Memoirs*, pp. 299–301; Ward, *War of the Revolution*, 2:717–30, 782, 868–69; Ltr, Clinton to Cornwallis, 11 Jun 1781, P.R.O.30/11/68, Cornwallis Papers; Rhys Isaac, *The Transformation of Virginia, 1740–1790* (Chapel Hill: University of North Carolina Press, 1982), pp. 275–76.

[23] Ltrs, James Madison to Edmund Pendleton, 16 Jan 1781, in *The Writings of James Madison*, ed. Galliard Hunt, 9 vols. (New York: G. P. Putnam's Sons, 1900), 1:122, and George Mason to George Mason Jr., 3 Jun 1781, in *The Papers of George Mason 1725–1792*, ed. Robert A. Rutland, 3 vols. (Chapel Hill: University of North Carolina Press, 1970), 2:693; Allan C. Fischer Jr., "My Chesapeake—Queen of Bays," *National Geographic* 158 (October 1980): 431.

[24] Ltr, Cornwallis to Clinton, 10 Apr 1781, in Ross, *Correspondence of Cornwallis*, 1:86–87.

Petersburg, south of Richmond, where he joined forces with a small British army commanded by Brig. Gen. Benedict Arnold. Arnold, the famed American traitor, had opened operations in Virginia by raiding up the James River in January 1781, and his activities highlighted the Old Dominion's vulnerability to amphibious operations. Maj. Gen. William Phillips had joined Arnold a few months later with 2,000 reinforcements and assumed command of the combined force, only to die of typhoid fever at Petersburg a week before Cornwallis' arrival. After Cornwallis absorbed Phillips' expedition, he had 8,000 seasoned regulars at his disposal, and he proceeded to subject Virginia to the ravages of war.[25] Two weeks after the junction of Cornwallis' and Phillips' forces, Virginian George Mason, a gentleman lawyer and a firm adherent of the Rebel cause, wrote in near despair:

> Our Affairs have been, for some time, growing from bad to worse. The Enemy's Fleet commands our Rivers, & puts it in their Power to remove their Troops from place to place, when and where they please without Opposition; so that we no sooner collect a Force sufficient to counteract them in one Part of the Country, but they shift to another, ravaging, plundering, and destroying everything before them. . . . The Enemy's capital Object, at this time, seems to be Virginia.[26]

For the next four months, Cornwallis terrorized the Patriots of Virginia with a new brand of war. One by one, he eliminated the mistaken assumptions that had hobbled the king's forces for the past six years. In their place, he introduced a simple but brutal strategy that strained Virginia's devotion to the cause of liberty. Less than a month after Cornwallis entered the Old Dominion, Richard Henry Lee, who as a delegate to the Continental Congress in 1776 had been one of the leaders in the drive to declare American independence, was sounding like a defeatist: "We shall receive all the injury before aid is sent to us—What will become of these . . . parts heaven knows—We and our property here are now within the power of the enemy." To that gloomy assessment, Lee added: "Cornwallis is the Scourge—& a severe one he is—The doings of

---

[25] Audrey Wallace, *Benedict Arnold: Misunderstood Hero?* (Shippensburg, Pa.: Burd Street Press, 2003), pp. 75–80; Johann von Ewald, *Diary of the American War: A Hessian Journal*, trans. and ed. Joseph P. Tustin (New Haven: Yale University Press, 1979), pp. 255–58, 266–75, 302; John Graves Simcoe, *Simcoe's Military Journal: A Military History of the Operations of a Partisan Corps, Called the Queen's Rangers* (1844; repr., New York: New York Times and Arno Press, 1968), pp. 158–72, 193–202; Rayford W. Logan, ed., "Memoirs of a Monticello Slave: As Dictated to Charles Campbell in the 1840s by Isaac, One of Thomas Jefferson's Slaves," *William and Mary Quarterly*, 3d ser., 8 (October 1951): 561–65; Harry M. Ward, *Duty, Honor or Country: General George Weedon and the American Revolution* (Philadelphia: American Philosophical Society, 1979), pp. 162–68; Ltr, Cornwallis to Clinton, 20 May 1781, in Ross, *Correspondence of Cornwallis*, 1:100; Lee, *Memoirs*, p. 413; Ltr, Arnold to Clinton, 12 May 1781, quoted in Tarleton, *Campaigns*, pp. 334–40; Stedman, *American War*, 2:354–55.

[26] Ltr, George Mason to George Mason Jr., 3 Jun 1781, in Rutland, *Papers of George Mason*, 2:693.

more than a year in the South are undoing very fast, whilst they rush to throw ruin into the other parts."[27]

One of Cornwallis' most striking tactical departures was to cease putting his trust in the Loyalists. He no longer wasted his time courting unreliable allies. All he asked of those white Virginians who claimed to support George III was that they stay out of his way.[28]

This public warning, which Cornwallis posted in the waning days of his Virginia campaign, characterized his new approach:

> The Inhabitants of Elizabeth City, York & Warwick Counties, being in the power of His Majesty's Troops, are hereby ordered to repair to Head Quarters at York Town on or before the 20th day of Aug[st] to deliver up their Arms, and to give their Paroles, that they will not in future take any part against His Majesty's Interest. And they are likewise directed to bring to Market the Provisions that they can spare, for which they will be paid reasonable prices in ready money. And notice is hereby given, that those who fail in complying with this Order will be imprisoned when taken, & their Corn and Cattle will be seized for the use of the Troops.[29]

Unlike other British commanders, Cornwallis kept his army on the move almost constantly. He did not just take cities and sit in them. "From the experience I have had," the earl reflected, "and the dangers I have undergone, one maxim appears to me to be absolutely necessary for the safe and honourable conduct of this war, which is,—that we should have as few posts as possible, and that wherever the King's troops are, they should be in respectable force."[30] By dint of frequent and rapid marches, Cornwallis kept the Rebels off-balance. He left his enemies no sanctuaries where they could rally or stockpile arms.[31]

Cornwallis also made certain that Virginia's civilians paid for their allegiance to the rebellion by suffering the horrors of war. He not only struck at the state's military capacity, but also at its citizens' purses. If Virginians wanted to defy royal authority, they would pay dearly for it. Cornwallis had his far-ranging army destroy anything that might be of use to the Patriot war

---

[27] Ltr, Richard Henry Lee to Arthur Lee, 4 Jun 1781, in Ballagh, *Letters of Richard Henry Lee*, 2:229–30.

[28] Ltr, St. George Tucker to Fanny Tucker, 11 Jul 1781, in Charles Washington Coleman Jr., "The Southern Campaign 1781 from Guilford Court House to the Siege of York Narrated by St. George Tucker in Letters to His Wife, Part II, The Peninsula Campaign," *Magazine of American History* 7 (September 1881): 207; Ltr, Thomas Nelson to Washington, 27 Jul 1781, in Palmer, *Calendar of Virginia State Papers*, 2:259; Tarleton, *Campaigns*, p. 296.

[29] Cornwallis, Notification, 9 Aug 1781, P.R.O. 30/11/101, Cornwallis Papers.

[30] Ltr, Cornwallis to Clinton, 26 May 1781, in Ross, *Correspondence of Cornwallis*, 1:100.

[31] Ltr, Cornwallis to Tarleton, 11 Jun 1781, P.R.O. 30/1/87, Cornwallis Papers.

effort—including private property. The following order, which the earl issued to his cavalry, typified his new strategy:

> All public stores of corn and provisions are to be burnt, and if there should be a quantity of provisions or corn collected at a private house, I would have you destroy it. . . . As there is the greatest reason to apprehend that such provisions will be ultimately appropriated by the enemy to the use of General Greene's army, which, from the present state of the Carolinas, must depend on this province for its supplies.[32]

Lt. Col. Banastre Tarleton, the commander of Cornwallis' cavalry, believed that terrorizing the inhabitants of Rebel districts was a "point of duty." He boasted that he would "carry the sword and fire through the Land." Everywhere they went, Cornwallis' soldiers promised to retaliate against the homes and persons of any Virginians who bore arms against the king. The property of those who figured prominently in the rebellion suffered thorough destruction.[33] This was how Thomas Jefferson, then Virginia's governor, described what Cornwallis did to his estate at Elkhill:

> He destroyed all my growing crops of corn and tobacco, he burned all my barns containing the same articles of the last year, having first taken what corn he wanted, he used . . . all my stocks of cattle, sheep, and hogs for the sustenance of his army, and carried off all the horses capable of service: of those too young for service he cut the throats, and he burnt all the fences on the plantation, so as to leave it an absolute waste.[34]

"This Family has not yet lost any Tobo [tobacco], Slaves, or other Property, by the Enemy," George Mason reassured his son on 3 June 1781, "but we are in daily expectation of sharing the same Fate with our Neighbors upon this, & the other Rivers; where many Familys have been suddenly reduced from Opulence to Indigence, particularly upon James River; the Enemy taking all the Slaves, Horses, Cattle, Furniture, & other Property, they can lay their Hands on."[35]

---

[32] Ltr, Cornwallis to Tarleton, 8 Jul 1781, quoted in Tarleton, *Campaigns*, pp. 402–03.

[33] Ltr, Pendleton to Madison, 30 Apr 1781, in *The Letters and Papers of Edmund Pendleton, 1734–1803*, ed. David John Mays, 2 vols. (Charlottesville: University of Virginia Press, 1967), 1:354; Richard K. MacMaster, ed., "News of the Yorktown Campaign: The Journal of Dr. Robert Honyman, April 17–November 25, 1781," *Virginia Magazine of History and Biography* 79 (October 1971): 400–401, 404, 411; Ltr, Richard Henry Lee to Arthur Lee, 4 Jun 1781, in Ballagh, *Letters of Richard Henry Lee*, 2:229–30. First quote from Tarleton, *Campaigns*, pp. 100–101. Simcoe, *Military Journal*, pp. 213–23. Second quote from Ltr, David Garland to Thomas Nelson, 23 Jul 1781, in Palmer, *Calendar of Virginia State Papers*, 2:240–41.

[34] Ltr, Jefferson to William Gordon, 16 Jul 1788, in Boyd, *Papers of Thomas Jefferson*, 13:363.

[35] Ltr, George Mason to George Mason Jr., 3 Jun 1781, in Rutland, *Papers of George Mason*, 2:690.

While threatening Virginia Rebels with instant impoverishment, Cornwallis kept the Americans from wearing down his troops with guerrilla warfare by making his army more mobile than Patriot forces. The earl's command was well suited for a war of swift maneuver. According to Sir Henry Clinton, "the chief part" of the royal troops in Virginia comprised "the elite of my army." Most of Cornwallis' British regiments had been campaigning in North America since 1775 and 1776, and they included such renowned formations as the Brigade of Foot Guards, the 23d Royal Welch Fusiliers, the 33d Foot (Cornwallis' own regiment), and the 71st Fraser's Highlanders. Long hours of drill and frequent combat experience left these regulars equally adept at the formal European tactics of the day and the open-order woodland skirmishing favored by Rebel irregulars. Among the most valuable units serving with Cornwallis were two green-coated Loyalist corps, the British Legion and the Queen's Rangers. The British Legion was something of a miniature army. Half of its members were cavalry and the other half infantry. The Legion followed a ruthless young Englishman named Banastre Tarleton. This hard-riding light dragoon reportedly indulged a taste for cruelty. Rebels claimed that Tarleton ordered his men to murder prisoners, and the Legion also possessed an unenviable reputation for looting. Like the British Legion, the Queen's Rangers was a composite organization. Close to 40 percent of the men were horse soldiers—hussars and light dragoons—while the rest were superbly conditioned light infantry. The leader of the Queen's Rangers was another alert and active young officer from England, Lt. Col. John Graves Simcoe. A master of partisan warfare, Simcoe delighted in luring his adversaries into cleverly laid ambushes. Nevertheless, he seems to have been cut from a different cloth than the impetuous Tarleton. Simcoe fought hard, but he had no stomach for atrocities. He effectively prevented the Queen's Rangers from molesting helpless prisoners and noncombatants.[36]

---

[36] Gregory J. W. Urwin, "The British Guards in the Revolution," *Campaigns* 6 (July/August 1981): 48–51; Robert D. Bass, *The Green Dragoon: The Lives of Banastre Tarleton and Mary Robinson* (1957; repr., Columbia, S.C.: Sandlapper Press, 1973), pp. 44–58, 72–126, 139–73; Philip R. N. Katcher, *Encyclopedia of British, Provincial, and German Army Units, 1775–1783* (Harrisburg, Pa.: Stackpole Books, 1973), pp. 1–148; idem, *The American Provincial Corps, 1775–1784* (Reading, England: Osprey Publishing, 1973), pp. 3–32; Isaac Samuel Harrell, *Loyalism in Virginia: Chapters in the Economic History of the Revolution* (Durham, N.C.: Duke University, 1926), p. 49; Tarleton, *Campaigns*, pp. 101, 270–76; Stedman, *American War*, 2:183; Lamb, *Journal*, p. 381. Quotes from Clinton, *American Rebellion*, p. 304. Simcoe, *Military Journal*, pp. 150–58, 165–67, 212; *Royal Gazette* (New York), 19 May, 14, 18, 25 Jul 1781; Ewald, *Diary*, pp. 140–50, 267–76, 278–80, 294, 296, 302; Lee, *Memoirs*, p. 413; Ltr, Simcoe to Cornwallis, 2 Jun 1781, P.R.O. 30/11/6, Cornwallis Papers. For discussions of the British Army's adoption of open-order tactics during the War of Independence, see Lawrence E. Babits, *A Devil of a Whipping: The Battle of Cowpens* (Chapel Hill: University of North Carolina Press, 1998), pp. 17–18; J. A. Houlding, *Fit for Service: The Training of the British Army, 1715–1795* (Oxford: Clarendon Press, 1981), pp. 237–38, 336–37. Colonel Tarleton's controversial American career is reassessed in Anthony J. Scotti, *Brutal Virtue: The Myth and Reality of Banastre Tarleton* (Bowie, Md.: Heritage Books, 2002).

By combining the mounted detachments from the British Legion and the Queen's Rangers, Cornwallis could count on the services of roughly five hundred hussars and light dragoons. That was the largest number of horsemen ever assembled by the British during the war in the South. The size of the earl's cavalry had a particularly intimidating effect on the Virginia militia.[37] Recognizing the enemy's superiority in mounted troops gave Cornwallis a pronounced advantage, the Marquis de Lafayette (the young French general commanding the Continental forces charged with the defense of Virginia) complained in a letter to General Washington:

> Was I to fight a battle I'll be cut to pieces, the militia dispersed, and the arms lost. Was I to decline fighting the country would think herself given up. I am therefore determined to scarmish, but not to engage too far, and particularly to take care against their immense and excellent body of horse whom the militia fears like they would so many wild beasts.[38]

Even as Lafayette wrote those words, however, Cornwallis took steps that prevented the Rebels from impeding the progress of British forces in Virginia. Since the late seventeenth century, the favorite hobbies of Virginia's gentry were breeding and racing fine horses. There was hardly a plantation in the Old Dominion that did not boast of a well-stocked stable full of thoroughbreds. When Cornwallis invaded Virginia, he seized these spirited chargers for his own use. Thanks to this inexhaustible supply of remounts, the earl's 500 light dragoons and hussars could travel thirty to seventy miles a day, which greatly increased the range and unsettling impact of their raids. Cornwallis also put 700 to 800 of his infantrymen on horseback, thus more than doubling his mounted strength.[39] On 4 June 1781, a worried Richard Henry Lee told

---

[37] MacMaster, "Journal of Robert Honyman," pp. 398, 403, 406; Lee, *Memoirs*, p. 415. Ltrs, Lafayette to Jefferson, 26 May 1781, in Boyd, *Papers of Thomas Jefferson*, 6:18; Richard Henry Lee to Washington, 12 Jun 1781, in Ballagh, *Letters of Richard Henry Lee*, 2:235; and H. Young to William Davies, 26 May 1781, in Palmer, *Calendar of Virginia State Papers*, 2:120. "State of the Troops that Marched with the Army under the Command of Lieut. Genl. Earl Cornwallis," 1 May 1781, P.R.O. 30/11/103, Cornwallis Papers; Scotti, *Brutal Virtue*, p. 41.

[38] Ltr, Lafayette to Washington, 24 May 1781, in *The Letters of Lafayette to Washington, 1777–1799,* ed. Louis Gottschalk, 2d ed. (Philadelphia: American Philosophical Society, 1976), p. 198.

[39] During the height of the Virginia campaign, the Queen's Rangers counted an enlisted strength of 632 men. Two hundred forty-eight of this number belonged to the regiment's four troops of hussars and light dragoons. The remaining 384 were divided among eleven infantry companies, which included one company armed with rifles and swords. While in Virginia, the rifle company usually operated as mounted infantry. In an engagement at Spencer's Ordinary on 26 June 1781, however, the versatile mounted riflemen joined the regiment's cavalry in delivering a horseback charge. Ludwig, Baron von Closen, *Revolutionary Journal, 1780–1783,* trans. and ed. Evelyn M. Acomb (Chapel Hill: University of North Carolina Press, 1958), pp. 176–77; Howard C. Rice and Anne S. K. Brown, trans. and eds., *The American Campaigns of Rochambeau's Army 1780, 1781, 1782, 1783*, 2 vols. (Princeton: Princeton University Press, 1972), 1:68; T. H. Breen, "Horses and Gentlemen: The Cultural Significance of Gambling

his brother, "The fine horses on the James river have furnished them with a numerous and powerful Cavalry."[40] British ships visiting Virginia brought exaggerated accounts of Cornwallis' enhanced mobility to New York. As the *Royal Gazette*, a Loyalist newspaper, informed its readers on 13 June, "By the fleet from Virginia we learn, that Lord Cornwallis's army is at Richmond . . . in excellent condition for service, and has lately been supplied with a great number of good horses, so that the army . . . produces from two to three thousand well mounted cavaliers." Another report in the same paper claimed "that his Lordship's whole army is now mounted, acting with great rapidity and decision."[41]

Hyperbole aside, the thing to remember is that Cornwallis had created a British army that could outrun its Rebel opponents for the first time in the American Revolution. Lafayette possessed only 4,500 frightened troops, many of them untrained, to counter Cornwallis' movements. That figure included no more than three hundred cavalry. To avoid encirclement or surprise by the earl's larger and faster army, Lafayette felt compelled to keep at least twenty to thirty miles away from the British. At that distance, he could neither oppose nor harass the Redcoats.[42] "The British have so many Dragoons," Lafayette curtly informed Governor Jefferson, "that it becomes impossible to stop or reconnoitre their movements."[43]

All through the spring and summer of 1781, Cornwallis found himself free to go where he wanted. Since Lafayette stayed out of harm's way, the earl kept his army intact and potent. He did not have to fight any bloody battles to advance his strategy. The ravaging of the Old Dominion proceeded unchecked. "The fact is," Richard Henry Lee related, "the enemy by a quick collection

---

among the Gentry of Virginia," *William and Mary Quarterly*, 3d ser., 34 (April 1977): 239–57; Isaac, *Transformation of Virginia*, pp. 98–101; MacMaster, "Journal of Robert Honyman," pp. 395, 399; Ltr, Lafayette to Jefferson, 28 May 1781, in Boyd, *Papers of Thomas Jefferson*, 6:26; Lee, *Memoirs*, pp. 303, 415; Tarleton, *Campaigns*, pp. 286, 288, 299; Lamb, *Journal*, p. 371; *Royal Gazette* (New York), 7 Jul 1781; "State of the Queen's Rangers 25th Aug.[t] 1781, With Casualties from 25th June to 24 Aug.[st] 1781"; "State of the Queen's Rangers 25th December 1781 with broken periods from 25 Oct. to 24 Dec.[r] 1781," both in Ward-Chipman Papers: Muster Master General's Papers and Regimental Papers, 1776–1785, Canadian National Archives, Ottawa, microfilm at the David Library of the American Revolution; Simcoe, *Military Journal*, pp. 107–08, 230, 233; De Witt Bailey, *British Military Flintlock Rifles, 1740–1840* (Lincoln, R.I.: Andrew Mowbray Publishers, 2002), pp. 72–74.

[40] Ltr, Richard Henry Lee to Arthur Lee, 4 Jun 1781, in Ballagh, *Letters of Richard Henry Lee*, 2:231.

[41] *Royal Gazette* (New York), 13, 27 Jun, 27 Jul 1781.

[42] William Feltman, *The Journal of Lieut. William Feltman, of the First Pennsylvania Regiment, 1781–82: Including the March into Virginia and the Siege of Yorktown* (1853; repr., New York: New York Times and Arno Press, 1969), pp. 2–6; Ltr, George Mason to George Mason Jr., 3 Jun 1781, in Rutland, *Papers of George Mason*, 2:693; Lee, *Memoirs*, pp. 414–15; Ward, *War of the Revolution*, 2:874–75; Ltr, Lafayette to Jefferson, 26 May 1781, in Boyd, *Papers of Thomas Jefferson*, 6:18; *Royal Gazette* (New York), 13, 20 Jun 1781.

[43] Ltr, Lafayette to Jefferson, 28 May 1781, in Boyd, *Papers of Thomas Jefferson*, 6:26.

of their force, & by rapid movements, are now in the center of Virginia, with an army of regular infantry greater than that of the compounded regulars and militia commanded by the Marquis [de Lafayette] & with 5 or 600 excellent cavalry . . . this Country is, in the moment of its greatest danger . . . abandoned to the Arts & the Arms of the Enemy."[44]

Although Cornwallis sought to subdue Virginia by striking at its civilian population, he did not allow his army to degenerate into a mob of freebooters. His war on private property proceeded under strict supervision. From Cole's Plantation, the earl admonished his army on 5 June 1781, "All private foraging is again For bid, and the out posts are not to Suffer any foraging party to pass without a Commissioned Officer." Six days earlier, the commander of the 43d Regiment of Foot announced, "Any Soldier absent from Camp without leave in writing from the Officer Commanding his Company will be punished as a Maroader."[45] Cornwallis also issued detailed regulations to govern the confiscation of civilians' horses.

> Commanding Officers of Corps are desired to prevent the scandalous practice of taking Horses from the Country people; when the Commanding Officers of Cavalry find any Horses suitable to their Service they will report their [having] taken them the next morning at Head Quarters Unless when they are detached; In which case the Report is to be made the Morning after their joining the Army. Receipts are to be given to Friends and Certificates to all doubtfull Persons; to be hereafter paid or not, according to their past and future Conduct, who are neither in Arms or public Employment, or have abandoned their Plantations.[46]

Those Redcoats and Loyalists who defied the earl's efforts to maintain discipline and order risked swift and merciless punishment. On 2 June 1781, Colonel Simcoe informed Cornwallis that two light dragoon privates from the Queen's Rangers had raped and robbed a woman named Jane Dickinson. After an inquiry established the two Loyalists' guilt, the earl directed that they be executed the following day. Four days later, Cornwallis required a field officer and a captain from each of his brigades, along with a junior officer and twenty men from each regiment, to witness the evening execution of a deserter from the 23d Royal Welch Fusiliers and two others from the 76th Regiment of Foot.[47]

---

[44] Ltr, Richard Henry Lee to James Lovell, 12 Jun 1781, in Ballagh, *Letters of Richard Henry Lee*, 2:237.

[45] Brigade Orders, 31 May 1781; General Orders, 5 Jun 1781; Regimental Orders, 31 May 1781. All in Orderly Book: H. B. M. 43d Regiment of Foot General Orders: From 23 May to 25 Aug 1781, Additional Ms. 42,449, British Library, London.

[46] Orders, 28 May 1781, in ibid.

[47] Ltr, Simcoe to Cornwallis, 2 Jun 1781, P.R.O. 30/11/6, Cornwallis Papers; Simcoe, *Military Journal*, p. 212; Orders, 7 Jun 1781, Orderly Book: H. B. M. 43d Regiment of Foot General Orders.

Cornwallis also attempted to restrain the depredations of British forces not under his personal command. Shortly after his arrival in Virginia, he complained to Clinton about "the horrid enormities which are committed by our Privateers in Chesapeak Bay." Appalled at plundering that served no military purpose, the earl beseeched his commander-in-chief, "I must join my earnest wish that some remedy may be applied to an evil which is so very prejudicial to His Majesty's Service."[48]

Cornwallis not only strove to prevent his new strategy from reaching inhumane extremes, but he also made guarded use of conciliatory gestures. On 14 August, he instructed one of his subordinates: "All Militia Men Prisoners of War taken before the 18th of June are to be released on parole, unless some particular Crime is alledged against them. I would have you detain all prisoners charged with heinous Offenses, & the very violent people of Princess Ann [County] & the Neighbourhood of Portsmouth who may be some security to those who have been favorable to us." Such magnanimity was lost on many of the earl's enemies, who were more impressed by the destructive impact that his army had on the areas it traversed. "Cornwallis' campaign and Tarleton's patrols ravaged the countryside, and destroyed the fields of maize to an extent where even inhabitants had scarcely enough for their subsistence," reported a French officer. "There is no hay at all in Virginia." An apprehensive gentleman living in Hampton County exclaimed, "Many persons in Virginia, with large fortunes, are totally ruined. The inhabitants in our county have not yet suffered much . . . but I fear the time of our distress is drawing near." After the British briefly occupied Williamsburg, a disconsolate major in the state militia wrote his wife, "Here they remained for some days, and with them pestilence and famine took root, and poverty brought up the rear. . . . As the British plundered all that they could, you will conceive how great an appearance of wretchedness this place must exhibit."[49]

As far as the white citizens of Virginia were concerned, however, the most unnerving thing Cornwallis did was to liberate their black slaves. Virginia's 200,000 bondmen made up 40 percent of the state's population. Had Cornwallis been permitted to follow his own instincts, these exploited masses might have tipped the balance in favor of his attempted conquest of the Old Dominion.[50]

In this politically correct era, most American history textbooks are sure to mention those African Americans who supported the Patriot cause. As Ellen Gibson Wilson has pointed out, however, "there has been some reluctance to face the implications of the fact that the overwhelming majority of blacks

---

[48] Ltr, Cornwallis to Clinton, 26 May 1781, P.R.O. 30/11/74, Cornwallis Papers.

[49] First quote from Ltr, Cornwallis to Brig Gen Charles O'Hara, 14 Aug 1781, P.R.O. 30/11/89, Cornwallis Papers. Second quote from Closen, *Revolutionary Journal*, p. 174. Third quote from *Royal Gazette* (New York), 29 Aug 1781. Fourth quote from Ltr, St. George Tucker to Fanny Tucker, 11 Jul 1781, in Coleman, "The Southern Campaign, 1781," p. 207.

[50] Ellen Gibson Wilson, *The Loyal Blacks* (New York: Capricorn Books, 1976), p. 25.

who acted from choice were pro-British." Historian David Waldstreicher put it more objectively when he said: "One of the less-well-known facts about the Revolutionary War is that African Americans fought on both sides, primarily with their own freedom in mind." [51] Statistics reveal that many African Americans harbored no loyalty to a movement that promised life, liberty, and the pursuit of happiness solely to white adult males. Of the 500,000 blacks who inhabited the thirteen colonies during the War of Independence, as many as 80,000 to 100,000 flocked to the king's forces.[52] Their reason was simple but compelling. As Rev. Henry Muhlenberg, a Lutheran minister who worked near Philadelphia, confided to his diary, blacks "secretly wished that the British army might win, for then all Negro slaves will gain their freedom." "It is said," Muhlenberg later observed, "that this sentiment is almost universal among the Negroes in America."[53]

The British did offer freedom of sorts to slaves who reached royal lines—provided the fugitives' owners were Rebels. That qualification was forgotten, however, as the news worked its way through the slave grapevine. Most blacks came to equate the sight of a soldier in a red coat with liberty.[54]

The British did not begin to suspect how far and wide this misconception had spread until they invaded the South, where the overwhelming number of slaves resided.[55] Dwelling upon his experiences in South Carolina, Colonel Tarleton reported "that all the negroes, men, women, and children, upon the approach of any detachment of the King's troops, thought themselves absolved from all respect to their American masters, and entirely released from servitude: Influenced by this idea, they quitted the plantations, and followed the army."[56]

---

[51] Ibid., p. 21; David Waldstreicher, *Runaway America: Benjamin Franklin, Slavery, and the American Revolution* (New York: Hill and Wang, 2004), p. 210.

[52] Benjamin Quarles, *The Negro in the American Revolution* (Chapel Hill: University of North Carolina Press, 1961), pp. 40, 51, 71–72, 116, 119; Richard B. Morris, *The American Revolution Reconsidered* (New York: Harper Torchbooks, 1967), p. 76; Benjamin Quarles, "The Negro Response: Evacuation with the British," in *Black History, a Reappraisal*, ed. Melvin Drimmer (Garden City, N.Y.: Doubleday, 1968), p. 133; Wilson, *Loyal Blacks*, pp. 21–22.

[53] Theodore G. Tappert and John W. Doberstein, trans., *The Journals of Henry Melchior Muhlenberg*, 3 vols. (Philadelphia: Muhlenberg Press, 1942–58), 3:78.

[54] *Royal Gazette* (New York), 25 Sep 1779; Sylvia R. Frey, "The British and the Black: A New Perspective," *Historian* 38 (February 1976): 225–29; Charles Francis Adams, ed., *The Works of John Adams*, 10 vols. (Boston: Little, Brown, 1850–56), 2:428; Wilson, *Loyal Blacks*, p. 29. Judith L. Van Buskirk offers a vivid discussion of the African Americans who fled to British-occupied New York in *Generous Enemies: Patriots and Loyalists in Revolutionary New York* (Philadelphia: University of Pennsylvania Press, 2002), pp. 129–54.

[55] David Ramsay, *The History of the Revolution of South-Carolina, from a British Province to an Independent State*, 2 vols. (Trenton: Isaac Collins, 1785), 2:31–32; Wilson, *Southern Strategy*, p. 118. The most comprehensive discussion of this issue is Sylvia R. Frey, *Water from the Rock: Black Resistance in a Revolutionary Age* (Princeton: Princeton University Press, 1991).

[56] Tarleton, *Campaigns*, p. 89.

As long as the British sought to win the allegiance of white Americans, they discouraged this black exodus. A few weeks before Clinton sailed from Charleston to New York, he instructed Cornwallis, "As to the Negroes, I will leave such orders as I hope will prevent the Confusion that would arise from a further desertion of them to us, and I will consider some Scheme of placing those We have on abandoned Plantations on which they may subsist. In the meantime Your Lordship can make such Arrangements as will discourage their joining us." The Redcoats even returned runaways to masters who were reputedly loyal or neutral. By the time Cornwallis entered Virginia, however, he no longer worried about the feelings of colonial slave owners, and he permitted black runaways to tag along with his soldiers.[57]

The response of Virginia's blacks astounded both the Patriots and the British. "The damage sustained by individuals on this occasion is inconceivable," testified Dr. Robert Honyman, a physician in Hanover County,

> especially in Negroes; the infatuation of these poor creatures was amazing: they flocked to the Enemy from all quarters, even from very remote parts. . . . Many Gentlemen lost 30, 40, 50, 60 or 70 Negroes beside their stocks of Cattle, Sheep & Horses. Some plantations were entirely cleared, & not a single Negro remained. Several endeavoured to bring their Negroes up the Country & some succeeded; but from others the slaves went off by the way & went to the Enemy.[58]

"Your neighbors Col. Taliaferro & Col. Travis lost every slave they had in the world," Richard Henry Lee informed his brother William, "and Mr. Paradise has lost all his but one—This has been the general case of all those who were near the enemy."[59] Other prominent Virginians told similar stories.[60] For instance, Thomas Nelson, the militia general who succeeded Jefferson as governor midway through Cornwallis' campaign, owned seven hundred slaves before the British entered Virginia. After Yorktown, no more than eighty to one hundred remained in his charge.[61]

---

[57] Frey, "British and the Black," pp. 228–29; Stedman, *American War*, 2:193, 217; Wilson, *Loyal Blacks*, pp. 32–36; Wilson, *Southern Strategy*, pp. 176, 234. Quotes from Ltr, Clinton to Cornwallis, 20 May 1780. Ltr, Nesbitt Balfour to Cornwallis, 24 Jun 1780. Last two in P.R.O. 30/11/2, Cornwallis Papers.

[58] MacMaster, "Journal of Robert Honyman," p. 394.

[59] Ltr, Richard Henry Lee to William Lee, 15 Jul 1781, in Ballagh, *Letters of Richard Henry Lee*, 2:242.

[60] Ltrs, George Mason to George Mason Jr., 3 Jun 1781, in Rutland, *Papers of George Mason*, 2:690; Pendleton to Madison, 30 Apr 1781, in Mays, *Letters of Edmund Pendleton*, 1:354; George Mason to Greene, 27 Jul 1781, quoted in Ward, *General George Weedon*, p. 205; Jefferson to William Jones, 5 Jan 1787, in Boyd, *Papers of Thomas Jefferson*, 11:16; and St. George Tucker to Fanny Tucker, 11 Jul 1781, in Coleman, "The Southern Campaign, 1781," p. 207.

[61] Closen, *Revolutionary Journal*, p. 180.

Cornwallis' soldiers actively encouraged Virginia slaves to follow them. Honyman, who refused to flee his home at the earl's approach, observed the enemy's recruitment practices. "Where ever they had an opportunity," Honyman confided to his journal, "the soldiers & inferior officers . . . enticed & flattered the Negroes, & prevailed on vast numbers to go along with them, but they did not compel any." Capt. Johann Ewald, the commander of a crack Hessian jaeger detachment with Cornwallis, explained his comrades' sudden passion for liberating slaves: "These people were given their freedom by the army because it was actually thought this would punish the rich, rebellious-minded inhabitants of . . . Virginia." Richard Henry Lee charged that "force, fraud, intrigue, theft, have all in turn been employed to delude these unhappy people [the slaves], and defraud their masters!" Despite such anguished assertions, there is abundant evidence that those slaves who joined the British did so freely. As one Virginia gentleman admitted, "Our negroes flock fast to them." Lafayette even reported to Washington that many of the Rebel commander-in-chief's slaves had joined the British.[62]

By the middle of June 1781, at least 12,000 runaway slaves were with Cornwallis' army. Jefferson later observed, "From an estimate I made at that time on the best information I could collect, I supposed the state of Virginia lost under Ld. Cornwallis's hands that year about 30,000 slaves."[63]

How all this appeared to the British is revealed in the diary of Captain Ewald:

> Every officer had four to six horses and three or four Negroes, as well as one or two Negresses for cook and maid. Every soldier's woman was mounted and also had a Negro and Negress on horseback for her servants. Each squad had one or two horses and Negroes, and every noncommissioned officer had two horses and one Negro. Yes, indeed, I can testify that every soldier had his Negro, who carried his provisions and bundles. This multitude always hunted at a gallop, and behind the baggage followed well over four thousand Negroes of both sexes and all ages. Any place this horde approached was eaten clean, like an acre invaded by a swarm of locusts.[64]

[62] First quote from MacMaster, "Journal of Robert Honyman," p. 400. Second quote from Ewald, *Diary*, p. 305. Third quote from Ltr, Richard Henry Lee to Washington, 17 Sep 1781, in Ballagh, *Letters of Richard Henry Lee,* 2:256. Logan, "Memoirs of a Slave," pp. 562–72; Ltr, Pendleton to Madison, 29 Jul 1782, in Mays, *Letters of Edmund Pendleton*, 2:402; "Jefferson's Statement of Losses to the British at His Cumberland Plantations in 1781," 27 Jan 1783, in Boyd, *Papers of Thomas Jefferson*, 6:224–25. Ltr, Lafayette to Washington, 23 Apr 1781, in Gottschalk, *Letters of Lafayette to Washington*, p. 187. Fourth quote from *Royal Gazette* (New York), 29 Aug 1781.

[63] Quote from Ltr, Jefferson to William Gordon, 16 Jul 1788, in Boyd, *Papers of Thomas Jefferson*, 13:364. Ewald, *Diary*, p. 305.

[64] Ewald, *Diary*, p. 305.

Virginia's fugitive slaves did more than serve the earl's soldiers as porters and body servants. The blacks also contributed substantially to Cornwallis' new style of warfare.

By encouraging the slaves to leave their masters, Cornwallis threatened Virginia with complete economic ruin. Slaves represented the currency whereby the Tidewater planters calculated their wealth. Slaves also provided the cheap labor undergirding the Old Dominion's agrarian prosperity. Thus Cornwallis robbed Virginia of the very means of production required to replace the vital resources his troops were destroying.[65]

The addition of thousands of African Americans to the British forces greatly augmented Cornwallis' ability to ravage the countryside. Dr. Honyman of Hanover County composed this vivid picture of one of Cornwallis' abandoned campsites:

> The day after the Enemy left Mrs. Nicholas's [plantation] I went over to her house, where I saw the devastation caused by the Enemy's encamping there, for they encamped in her plantation all round the house. The fences [were] pulled down & much of them burnt; Many cattle, hogs, sheep & poultry of all sorts killed; 150 barrels of corn eat up or wasted; & the offal of the cattle &c. with dead horses & pieces of flesh all in a putrefying state scattered over the plantation.[66]

Virginia's fugitive slaves also served Cornwallis in a more deliberate fashion. Runaways sometimes acted as spies and guides for the British. The blacks frequently showed their new friends where fleeing masters had hidden their valuables and livestock.[67] In fact, the African Americans delivered so many horses to Cornwallis that Lafayette exclaimed, "Nothing but a treaty of alliance with the negroes can find out dragoon horses, and it is by those means the ennemy have got a formidable cavalry."[68] At other times, the blacks provided manual labor for the British Army. As one Virginian put it, the fugitives "ease the soldiery of the labourer's work." A corps of "Negro Pioneers" (military laborers), originally formed by General Phillips, buried the offal from butchered cattle after Cornwallis' troops received issues of

---

[65] William Duane and Thomas Balch, trans. and eds., *The Journal of Claude Blanchard: Eyewitness Accounts of the American Revolution* (1876; repr., New York: New York Times and Arno Press, 1969), p. 162; Rice and Brown, *Rochambeau's Army*, 1:67; Closen, *Revolutionary Journal*, p. 187; Isaac, *Transformation of Virginia*, p. 136.

[66] Quote from MacMaster, "Journal of Robert Honyman," pp. 401–02. Ewald, *Diary*, p. 305; Closen, *Revolutionary Journal*, p. 174; Ltr, St. George Tucker to Fanny Tucker, 11 Jul 1781, in Coleman, "The Southern Campaign, 1781," pp. 207–08.

[67] Closen, *Revolutionary Journal*, p. 166; McMaster, "Journal of Robert Honyman," p. 399; Ltr, Lafayette to Jefferson, 28 May 1781, in Boyd, *Papers of Thomas Jefferson*, 6:26; Rice and Brown, *Rochambeau's Army*, 1:154; Tarleton, *Campaigns*, p. 353; Simcoe, *Military Journal*, p. 165.

[68] Ltr, Lafayette to Washington, 20 Jul 1781, in Gottschalk, *Letters of Lafayette to Washington*, p. 209.

fresh meat, thus eliminating a nauseating stench and also a health hazard. The black pioneers and officers' servants pulled double duty as stevedores whenever Cornwallis used ships to transport soldiers, equipment, and supplies. The extensive earthworks that Cornwallis had erected at Portsmouth and Yorktown were built largely by black muscle. Finally, the defection of so many slaves spread the fear of servile revolt—the white South's most dreaded nightmare—throughout Virginia.[69]

As much as Cornwallis benefited from the specter of black rebellion, he did not intend to unleash a racial reign of terror against the Old Dominion's white population. The earl composed numerous regulations throughout his Virginia campaign aimed at ensuring orderly conduct among slaves seeking his protection. To restore his army's proper military appearance and free his columns of unnecessary encumbrances, Cornwallis attempted to restrict the number of horses and blacks employed by his officers. A colonel, lieutenant colonel, or major of infantry was entitled to "5 Horses and 2 Negroes." A captain could have three horses and one black servant, regimental staff officers and subalterns could each have a pair of mounts and a single servant, and a surgeon was limited to one horse and one black. Sergeants major, the most senior noncommissioned officers in the earl's regiments, were also permitted one horse and one black servant apiece. Except for those detailed for mounted service, enlisted infantrymen did not receive permission to ride horses, and no one below the rank of sergeant major could enjoy the services of black servants. Cornwallis also stipulated, "No woman [white camp follower] or negro to possess a Horse, nor any negro to be Suffered to ride on a March except such as belong to publick departments."[70]

To distinguish the African Americans who were authorized to accompany the army's different units from those who were not, Cornwallis decreed on 21 May 1781, "The number or names of Corps to be marked in a Conspicuous manner on the Jacket of each negro." A week later, the earl informed his army, "All Negros who are not marked agreeable to the Orders repeated at Petersburg will be taken up and sent away from the Army."[71]

Cornwallis' headquarters frequently reminded unit commanders to purge their ranks of surplus horses and blacks. Typical of such orders was this one

---

[69] *Royal Gazette* (New York), 29 Aug 1781; Joseph G. Rosengarten, trans., "Popp's Journal, 1777–1783," *Pennsylvania Magazine of History and Biography* 26 (1902): 38; Ltr, Phillips to Clinton, 3 Apr 1781, P.R.O. 30/11/96, Cornwallis Papers. First quote from Regimental Orders, 28 Jul 1781. After Orders, 4 Aug 1781; Morning Regimental Orders, 25 Aug 1781. Last three in Orderly Book: H. B. M. 43d Regiment of Foot General Orders. Ltrs, Cornwallis to O'Hara, 4 Aug 1781, in Ross, *Correspondence of Cornwallis*, 1:113, and Pendleton to Madison, 10 Sep 1781, in Mays, *Letters of Edmund Pendleton*, 1:371; Robert J. Tilden, trans., "The Doehla Journal," *William and Mary Quarterly*, 2d ser., 22 (July 1942): 243.

[70] General Orders, 21 May 1781, in Orderly Book: H. B. M. 43d Regiment of Foot General Orders.

[71] General Orders, 21 May 1781, and Orders, 28 May 1871, both in ibid.

issued on 5 June: "Lord Cornwallis desires the Commanding officers of Corps to Examine Strictly what number of Negores there are with their respective Corps and See that no more are kept than those allowed by the regulation and They will order all the abel'd bodied Negroes which they find above their Number allowed to officers to be taken up and Sent to Capt Brown of the Pioneers."[72]

Some of Cornwallis' officers, sharing his sense of military decorum, conscientiously enforced their commander's orders. On 4 June, Maj. George Hewett, the commander of the 43d Regiment of Foot, warned his noncommissioned officers and privates: "Any Man found Guilty of sending the Negroes of the Regiment plundering or Maroding the smallest Article from the Houses of the Inhabitants will be severely punished." Captain Ewald, who joined Cornwallis on 21 June after recovering from a wounded leg, discovered that his jaeger detachment possessed more than twenty horses, and that "almost every jager had his Negro." With professional pride, Ewald scribbled in his diary, "But within twenty-four hours, I brought everything back on the track again." Ewald also noted, however, that in other units "this order was not strictly carried out," and "the greatest abuse arose from this arrangement." The no-nonsense Hessian officer blamed the situation on "the indulgent character of Lord Cornwallis."[73] In reality, the earl made repeated efforts to control his black camp followers and keep them from undermining his troops' discipline and the army's ability to respond to any threat.[74]

Although military expedience governed the earl's treatment of Virginia's slaves, he did betray a glimmer of sympathy for the runaways. In late July 1781, Thomas Nelson, Virginia's newly installed governor, sent Cornwallis a curious letter. "The frequent Applications that are made to me by the Citizens of this Commonwealth," Nelson wrote, "to grant Flags for the Recovery of their Negroes & other Property, taken by the Troops under your Command, induce me to address your Lordship for Information, whether Restitution will be made at all, what Species of Property will be restored, & who may expect to be the Object of such an Indulgence."[75]

Cornwallis replied with a polite but carefully worded note that must have given Nelson little satisfaction:

> No Negroes have been taken by the British Troops by my orders nor to my knowledge, but great numbers have come to us from different parts of the Country. Being desirous to grant any indulgence to individuals that I think

---

[72] Quote from General Orders, 5 Jun 1781, in ibid. See also Regimental Orders, 5 Jun 1781; Brigade Orders, 18 Jun 1781; General Orders, 25 Jun 1781. All in ibid.

[73] Regimental Orders, 4 Jun 1781, in ibid.; Ewald, *Diary*, pp. 305–06.

[74] Orders, 28 May 1781; Brigade Orders, 8 Jun 1781; General Orders, 11 Aug 1781; General Orders, 20 Aug 1781; General Orders, 25 Aug 1781. All in Orderly Book: H. B. M. 43d Regiment of Foot General Orders.

[75] Ltr, Thomas Nelson to Cornwallis, 23 Jul 1781, P.R.O. 30/11/90, Cornwallis Papers.

> consistent with my public duty, Any proprietor not in Arms against us, or holding an Office of trust under the Authority of Congress and willing to give his parole that he will not in future act against His Majesty's interest, will be indulged with permission to search the Camp for his Negroes & take them if they are willing to go with him.[76]

By the summer of 1781, Lord Cornwallis' new strategy of conquest bore a strong resemblance to the hard war policies that another invading army would adopt to pacify the American South eight decades later. In his own way, Cornwallis taught the Old Dominion the same lesson that Maj. Gens. William T. Sherman and Philip H. Sheridan would administer to the Confederacy during the Civil War. A century after Cornwallis' Virginia campaign, Sheridan captured the essence of that lesson in his memoirs: "Death is popularly considered the maximum of punishment in war, but it is not; reduction to poverty brings prayers for peace more surely and more quickly than does the destruction of human life, as the selfishness of man has demonstrated in more than one great conflict."[77]

Cornwallis' impromptu version of hard war was steadily forcing Virginia to its knees. The startling mobility of the earl's army denied local Continental forces the opportunity to engage in either conventional or guerrilla warfare. Cornwallis' policy of property despoliation also neutralized Virginia's last remaining line of defense, the militia. The strength and speed of British forces terrified Virginia's citizen-soldiers. Militiamen grew reluctant to take up arms lest they provoke the Redcoats into destroying their homes.[78] The militiamen also feared to leave their families alone with their slaves. "There were . . . forcible reasons which detained the militia at home," explained Edmund Randolph, who had been a Virginia delegate to Congress. "The helpless wives and children were at the mercy not only of the males among the slaves but of the very women, who could handle deadly weapons; and those could not have been left in safety in the absence of all authority of the masters and union among the neighbors."[79]

At this critical juncture, the swiftness of Cornwallis' movements made it impossible for Virginia's state government to function. On 3 June 1781, British

[76] Ltr, Cornwallis to Thomas Nelson, 6 Aug 1781, P.R.O. 30/11/90, Cornwallis Papers.

[77] Philip H. Sheridan, *Personal Memoirs of P. H. Sheridan*, 2 vols. (New York: Charles L. Webster, 1888), 1:487–88.

[78] Ltrs, Lafayette to Thomas Nelson, 26 Aug 1781, in *Lafayette in Virginia: Unpublished Letters*, ed. Gilbert Chinard (Baltimore, Md.: Johns Hopkins Press, 1928), p. 54; St. George Tucker to Fanny Tucker, 14 Sep 1781, in Coleman, "The Southern Campaign, 1781," pp. 206, 211; Richard Henry Lee to Washington, 12 Jun 1781, in Ballagh, *Letters of Richard Henry Lee*, 2:233; and Lafayette to Jefferson, 28 May 1781, in Boyd, *Papers of Thomas Jefferson*, 6:26. MacMaster, "Journal of Robert Honyman," pp. 393, 394, 416; Ward, *General George Weedon*, pp. 203, 216.

[79] Randolph, *History of Virginia*, p. 285. This fear was an American military weakness throughout the Revolution. As historian David K. Wilson observed, "The threat of a slave insurrection (and/or Indian attacks in the case of frontier counties) usually kept half of a southern county's militia at home." See Wilson, *Southern Strategy*, p. 3.

cavalry and mounted infantry raided the Virginia Assembly at Charlottesville, capturing seven legislators and forcing Governor Jefferson and the rest of the assemblymen to scatter for safety. "Lt. Colonel Tarleton took some Members of the Assembly at Charlottesville," Cornwallis boasted, "& destroyed there & on his return 1000 stand of good Arms, some Clothing & other Stores & between 4 & 500 barrels of Powder without opposition." In addition to Jefferson, many other well-known Virginians, including Richard Henry Lee and Edmund Pendleton, fled at the Redcoats' approach, depriving the Patriot cause of some of its best political leaders.[80]

Being denied protection by a skittish state government, lacking any hint of aid from the Continental Congress or America's French allies, and facing the prospect of economic disaster, the people of Virginia began to consider making peace with Great Britain. The inhabitants of Norfolk, Princess Anne, and Nansemond counties placed themselves under British protection. The men of Montgomery, Bedford, and Prince Edward counties ignored all summons for militia duty. When state officials tried to raise the militia in Accomack, Northampton, and Lancaster counties, they encountered opposition from armed mobs. Farmers living around the British base at Portsmouth started trading with the enemy, sometimes bringing the Redcoats intelligence about Rebel activities.[81] One of Cornwallis' Hessian corporals marveled at the Virginians' change of heart: "Toward us [the Portsmouth garrison] they were rather agreeable and showed more respect than in other provinces, especially the Virginia women had more affection for the Germans."[82] Defeatist sentiment reached such dangerous levels that Richard Henry Lee recommended that General Washington return to Virginia with his troops and assume dictatorial

---

[80] Emory G. Evans, *Thomas Nelson of Yorktown: Revolutionary Virginian* (Charlottesville: University Press of Virginia, 1975), pp. 100–101; Ltrs, Pendleton to Madison, 4 May 1781, and Pendleton to Madison, 6 Jul 1781, both in Mays, *Letters of Edmund Pendleton*, 1:356, 365; Lee, *Memoirs*, p. 438; Stedman, *American War*, 2:387–88; Tarleton, *Campaigns*, pp. 295–97; *Royal Gazette* (New York), 4 Jul 1781. Quote from Ltr, Cornwallis to Clinton, 30 Jun 1781, P.R.O. 30/11/74, Cornwallis Papers. Ltrs, Richard Henry Lee to Arthur Lee, 13 May 1781, in Ballagh, *Letters of Richard Henry Lee*, 2:233, and George Mason to Pearson Chapman, 31 May 1781, in Rutland, *Papers of George Mason*, 2:688. MacMaster, "Journal of Robert Honyman," p. 401.

[81] Ltrs, Richard Henry Lee to Arthur Lee, 4 Jun 1781, in Ballagh, *Letters of Richard Henry Lee*, 2:230, and George Mason to George Mason Jr., 3 Jun 1781, in Rutland, *Papers of George Mason*, 2:693–94; Tarleton, *Campaigns*, pp. 297–98. Ltrs, George Corbin to Jefferson, 31 May 1781, in Boyd, *Papers of Thomas Jefferson*, 6:44–46. James Arbuckle, Charles Bagwell, and D. Bowman to Jefferson, 15 May 1781; J. Parker to the Speaker of the Virginia Assembly, 9 Jun 1781; J. Parker to Thomas Nelson, 29 Jun 1781; Preston to Davies, 28 Jul 1781; Preston to Thomas Nelson, 28 Jul 1781; and James Innes to Thomas Nelson, 29 Jul 1781, last six in Palmer, *Calendar of Virginia State Papers*, 2:97–100, 151, 189, 246–67. Harrell, *Loyalism in Virginia*, pp. 59–62; Tilden, "Doehla Journal," p. 240; Isaac, *Transformation of Virginia*, p. 276.

[82] Tilden, "Doehla Journal," p. 240.

powers until the crisis passed.[83] Jefferson too urged Washington to hasten to the Old Dominion "to lend us Your personal aid."[84]

Although Cornwallis made Virginia howl, he failed to attain the objective that ostensibly drew him there in the first place—crippling General Greene's logistical system. To be sure, the earl's presence in the Old Dominion worried the Quaker general. As he reminded Lafayette on 9 June 1781, "Virginia is a capital link in the chain of communication and must not be left to sink under the oppression of such formidable attacks as are making upon her." Greene's concern for Virginia was compounded by the difficulty he experienced in supplying his army in South Carolina. "I can see no place where an Army of any considerable force can subsist for any length of time; and the horses are so destroyed in this Country that subsistence cannot be drawn from a distance," he had observed in May. With good reason, Lt. Col. Henry Lee, one of Greene's most active subordinates, flattered his commander, "I am also conscious that no General ever commanded troops worse appointed or worse supplyed, than those which form your present army." Cornwallis not only destroyed or confiscated resources that might have gone to Greene, but he also cost the latter considerable reinforcements. Greene had to halt the southward march of Brig. Gen. Anthony Wayne's Pennsylvania line and Continental recruits raised in Virginia to bolster Lafayette's numbers.[85]

Despite all these handicaps, Greene managed to keep his army alive. As Washington's former quartermaster general, he was probably the best-qualified officer in the Continental service to confront such a challenge. He purchased some of what he needed from various sources in North Carolina and made up the difference by living off the land. He sent out strong foraging parties to requisition dragoon horses, draft animals, edible livestock, and grain from Rebel and Loyalist farmers alike. Greene also appealed to South Carolina's partisan leaders—Brig. Gens. Thomas Sumter, Francis Marion, and Andrew Pickens—to send him some of the weapons, ammunition, and food that they had captured from the British. "I have Ten waggons on their way to you With Meal," Sumter informed Greene on 2 May 1781. In addition, a caravan of nearly two dozen wagons containing clothing and ammunition from the north managed to slip through Virginia before Cornwallis rendezvoused with Arnold at Petersburg. The passage of such convoys became increasingly difficult after the earl unleashed his wide-ranging cavalry and mounted infantry on the Old Dominion. Fortunately

---

[83] Ltrs, Richard Henry Lee to Washington, 12 Jun 1781, and Richard Henry Lee to James Lovell, 12 Jun 1781, both in Ballagh, *Letters of Richard Henry Lee*, 2:233–37.

[84] Quote from Ltr, Jefferson to George Washington, 28 May 1781, in Boyd, *Papers of Thomas Jefferson*, 6:33, and see also p. 32.

[85] Ltrs, Greene to Steuben, 2 Apr 1781; Lt Col Henry Lee to Greene, 2 Apr 1781; Greene to Samuel Huntington, 5 May 1781, 9 Jun 1781; and Greene to Lafayette, 9 Jun 1781, all in *The Papers of General Nathanael Greene*, ed. Richard K. Showman et al., 13 vols. (Chapel Hill: University of North Carolina Press, 1976–2005), 8:23, 28 (third quote), 206 (second quote)–07, 363, 366 (first quote)–67.

for the Rebels, British efforts to interdict the Virginia lifeline were short-lived. Interference from above brought a premature close to Cornwallis' campaign to knock the state out of the war.[86]

Cornwallis had plunged into Virginia without seeking permission from his immediate superior, Sir Henry Clinton. Clinton would later call that move "a measure . . . determined upon without my approbation, and very contrary to my wishes and intentions"—an opinion he made no effort to hide from his aristocratic subordinate at the time. Clinton faulted Cornwallis for exposing the Carolinas and Georgia to recapture by Greene. The British commander-in-chief also still clung to his faith in the Loyalists. He considered recalling a large number of the troops he had sent to the Chesapeake and using them instead to inspire an uprising in Maryland, Delaware, or southeastern Pennsylvania. Fear of a possible Franco-American siege of New York also made him contemplate a concentration of force there. At the same time, personal insecurity affected Clinton's strategic thinking. He and Cornwallis did not like each other, and they were rivals. Despite the heavy losses the earl suffered at Guilford Court House, his aggressive efforts to crush the rebellion contrasted sharply with Clinton's relative inactivity at New York. Suspecting that the earl's success might precipitate his own removal, Clinton brought an end to Virginia's agony. In the middle of the summer, he ordered Cornwallis to retire to the coast, set up a naval base, and send 2,000 troops back to New York. An exasperated Cornwallis began entrenching at Yorktown on the York River on 2 August 1781.[87]

---

[86] Ltrs, Sumter to Greene, 2 May 1781; Pickens to Greene, 25 May 1781; and Greene to Lt Col Henry Lee, 29 Jun 1781, all in *Papers of General Nathanael Greene*, 8:193 (quote), 310–11, 473. Ltrs, Marion to Greene, 21 Apr 1781, Nathanael Greene Papers, Henry E. Huntington Library, San Marino, Calif., and Davies to Greene, 17 Jun 1781, Nathanael Greene Papers, William L. Clements Library, University of Michigan, Ann Arbor; Lawrence E. Babits, "Supplying the Southern Continental Army, March 1780 to September 1781," *Military Collector & Historian: Journal of the Company of Military Historians* 47 (Winter 1995): 163–71; Theodore Thayer, *Nathanael Greene: Strategist of the American Revolution* (New York: Twayne Publishers, 1960), pp. 336–39, 349, 364; Terry Golway, *Washington's General: Nathanael Greene and the Triumph of the American Revolution* (New York: Henry Holt, 2005), pp. 264–66, 269, 271–72, 276, 278–79. Hugh F. Rankin, *Francis Marion: The Swamp Fox* (New York: Thomas Y. Crowell, 1973), pp. 184–238. For an insightful examination of the Continental Army's logistical system, which General Greene did so much to build, see E. Wayne Carp, *To Starve the Army at Pleasure: Continental Army Administration and American Political Culture 1775–1783* (Chapel Hill: University of North Carolina Press, 1984).

[87] Quote from Henry Clinton, *The Narrative of Lieutenant-General Sir Henry Clinton, K.B., relative to His Conduct during Part of His Command of the King's Troops in North America* (London: J. Debrett, 1783), p. 8, and see also p. 7; idem, *Observations on Some Parts of the Answer of Earl Cornwallis to Sir Henry Clinton's Narrative* (London: J. Debrett, 1783), pp. 11–12, 16. Ltrs, Clinton to Cornwallis, 8 Jun 1781; Clinton to Cornwallis, 11 Jun 1781; Clinton to Cornwallis, 15 Jun 1781; Clinton to Cornwallis, 19 Jun 1781; and Clinton to Cornwallis, 8 Jul 1781, last five in *The Campaign in Virginia 1781: An Exact Reprint of Six Rare Pamphlets on the Clinton-Cornwallis Controversy with Very Numerous Important Unpublished Manuscript Notes by Sir Henry Clinton, K.B.*, ed. Benjamin Franklin Stevens, 2 vols. (London: Privately Printed, 1888): 2:14–17, 18–23, 24–25, 26–28, 29–30, 49–56, 62–65. Ltrs, Cornwallis to Clinton, 30 Jun

Now fate turned against the British. At the end of August, a French fleet appeared off Chesapeake Bay, denying Cornwallis access to the sea. Seizing this opportunity, Washington pulled out of his lines around New York and slipped down to Virginia with a strong Franco-American army. By 28 September 1781, Cornwallis and his six thousand weary regulars found themselves besieged by nearly seventeen thousand Americans and Frenchmen.[88]

Cornwallis knew he was in a tight spot. Although he sympathized with the black runaways under his protection, he was the king's servant first. Hoping to stretch his army's provisions until Clinton could come to the rescue, the earl ordered all but 2,000 of the slaves sheltering at Yorktown expelled from British lines. Besides being terrified at the thought of returning to their vengeful masters, many of the cast-off blacks were seriously ill. They had contracted smallpox in the earl's camps. Frightened by what the future might bring and weakened by disease, hundreds of runaways simply lay down in the no-man's-land between the opposing trenches, where they died of exposure, illness, and starvation. The remainder took shelter in the woods around Yorktown. Few survived to witness Cornwallis' surrender on 19 October 1781.[89] Jefferson later

---

1781; Cornwallis to Clinton, 24 Jul 1781; and Cornwallis to Clinton, 22 Aug 1781, last three in Ross, *Correspondence of Cornwallis*, 1:103–06, 107–10, 112, 113–16, 117. Ltrs, Clinton to Cornwallis, 11 Jun 1781, P.R.O. 30/11/68, Cornwallis Papers; James Robertson to William Knox, 12 Jul 1781; Robertson to Lord Amherst, 8 Dec 1781; and Robertson to Amherst, 27 Dec 1781, last three in *The Twilight of British Rule in Revolutionary America: The New York Letter Book of General James Robertson, 1780–1783*, ed. Milton M. Klein and Ronald W. Howard (Cooperstown, N.Y.: New York State Historical Association, 1983), pp. 209, 231, 234. Ltr, John Ross to Lord Ankerville, 2 Mar 1781, RH15/44/103, Papers of John Ross, National Archives of Scotland, Edinburgh, Scotland; Ira D. Gruber, ed., *John Peebles' American War: The Diary of a Scottish Grenadier, 1776–1782* (Mechanicsburg, Pa.: Stackpole Books, 1998), p. 464; Don Higginbotham, *The War of American Independence: Military Attitudes, Policies, and Practice, 1763–1789* (New York: Macmillan, 1971), pp. 376–79; Clinton, *American Rebellion*, pp. 284, 289–90, 301–02; Mackesy, *War for America*, p. 409; Wickwire, *Cornwallis*, pp. 352–53; William Smith, *Historical Memoirs of William Smith, 1778–1783*, ed. William H. W. Sabine (New York: New York Times and Arno Press, 1971), pp. 406, 407, 418.

[88] Tarleton, *Campaigns*, pp. 368–69; Ward, *War of the Revolution*, 2:879–87; Stedman, *American War*, 2:414; Tappert and Doberstein, *Journals of Henry Muhlenberg*, 3:443; Donald Jackson, ed., *The Diaries of George Washington*, 6 vols. (Charlottesville: University of Virginia Press, 1976–1979), 3:409–10.

[89] John C. Dann, ed., *The Revolution Remembered: Eyewitness Accounts of the War for Independence* (Chicago: University of Chicago Press, 1980), p. 244; Joseph Plumb Martin, *Private Yankee Doodle: Being a Narrative of Some of the Adventures, Dangers, and Sufferings of a Revolutionary Soldier*, ed. George F. Scheer (Boston: Little, Brown, 1962), pp. 241–42; Fleming, *Beat the Last Drum*, pp. 175–76, 256; Ewald, *Diary*, pp. 314, 318, 328, 335–36; Edward M. Riley, "St George Tucker's Journal of the Siege of Yorktown, 1781," *William and Mary Quarterly*, 3d ser., 5 (July 1948): 387; MacMaster, "Journal of Robert Honyman," p. 420; Logan, "Memoirs of a Slave," p. 572; Feltman, *Journal*, p. 6; Frey, "British and the Black," pp. 232–33. Ltrs, Cornwallis to Clinton, 22 Aug 1781; Cornwallis to Clinton, 16 Sep 1781; and Cornwallis, 20 Oct 1781, last three in Ross, *Correspondence of Cornwallis*, 1:117, 120, 127–28. Wilson, *Loyal Blacks*, p. 42; Tilden, "Doehla Journal," pp. 238, 241–46, 251; Rosengarten, "Popp's Journal," p. 41. A medical history of Cornwallis' Virginia campaign that

claimed that 27,000 of the 30,000 fugitive slaves died of diseases brought to Virginia by the British.[90]

Cornwallis had received an inkling of the bleak future in store for his black allies months before he was trapped at Yorktown. Within weeks of Cornwallis' arrival in Virginia, the blacks following the British began exhibiting the unmistakable symptoms of smallpox. On 18 June 1781, the earl's headquarters advised the army's "Diferent deppertments who have Negroes in their employ to get them inoculeted." That same day, Lt. Col. Thomas Dundas, one of Cornwallis' brigade commanders, cautioned his officers: "Returns to be given in by the 43d, 76th, and 80th Regiments as Soon as possible of the number of Men in the Regiments who have not had the Small pox, and as a number of Negroes belonging to the Army now have the Small Pox, and a number going to be Invealeted, it is recommended to such men as never had Such Disorder to avoid Communicating with the negroes until such a proper opportunity shall be found to have them inoculated." Inoculations were administered to the troops, but it is not apparent that runaway slaves received the same treatment.[91]

What is clear is that smallpox was soon running rampant among those African Americans who were exposed to the earl's Redcoats, Hessians, and Loyalists. Lt. William Feltman of General Wayne's brigade of Pennsylvania Continentals, which shadowed British movements in late June, found the enemy's route of march littered with sick and abandoned blacks. He described

---

credits another origin for the smallpox plague that swept the blacks who followed the British is Philip Ranlet, "The British, Slaves, and Smallpox in Revolutionary Virginia," *Journal of Negro History* 84 (Summer 1999): 217–26.

[90] Ltr, Jefferson to William Gordon, 16 Jul 1788, in Boyd, *Papers of Thomas Jefferson*, 13:364. An untold number of runaway slaves escaped either death or recapture. When three transports arrived at Yorktown on 3 November 1781 to carry paroled British officers back to New York, some of them smuggled their black mistresses and servants on board. Other fugitives represented themselves as "freemen" and obtained employment among the victorious Continental and French officers. "Negroes without masters found new ones among the French," reported one French lieutenant, "and we garnered a veritable harvest of domestics. Those among us who had no servant were happy to find one so cheap." An exasperated General Washington repeatedly commanded his officers to surrender their newly acquired servants. In "After Orders" issued on 25 October, Washington decreed, "All Officers of the Allied Army and other persons of every denomination concerned are directed not to suffer any such negroes or mulattoes to be retained in their Service but on the contrary to cause them to be delivered to the Guards which will be establish'd for their reception at one of the Redoubts in York and another in Gloucester." Thus Washington immediately converted the army that won independence into a posse of slave catchers and created another one of the ironies that continue to trouble students of American history. First quote from Rice and Brown, *Rochambeau's Army*, 1:64. Second quote from John C. Fitzpatrick, ed., *Writings of George Washington*, 39 vols. (Washington, D.C.: Government Printing Office, 1931–1944), 23:265.

[91] First quote from General Orders, 18 Jun 1781. Second quote from Brigade Orders, 18 Jun 1781. Brigade Orders, 28 Jun 1781. All in Orderly Book: H. B. M. 43d Regiment of Foot General Orders. Ranlet, "British, Slaves, and Smallpox," p. 222. For more on the impact of smallpox on the Revolutionary War, see Elizabeth A. Fenn, *Pox Americana: The Great Smallpox Epidemic of 1775–82* (New York: Hill & Wang, 2001).

them as "starving and helpless, begging of us as we passed them for God's sake kill them, as they were in great pain and misery." Feltman accused the British of frequently leaving black smallpox victims lying in their wake "in order to prevent the Virginia militia from pursuing them."[92]

"Above 700 Negroes are come down the River in the Small Pox," Maj. Gen. Alexander Leslie, the commander of the British garrison at Portsmouth, wrote Cornwallis on 13 July 1781. Leslie was coldhearted enough to continue to use the stricken blacks as military assets. "I shall distribute them," he informed Cornwallis, "about the Rebels Meantimes."[93] After Cornwallis decided to concentrate his forces at Yorktown, he detailed Brig. Gen. Charles O'Hara to oversee the evacuation of Portsmouth. A warm and friendly Irishman, O'Hara sent his commander a heartrending report on the rapidly deteriorating condition of the post's black population.

> I shall continue till I receive Your positive instructions to the contrary, to victual the *Sick Negroes*, above *1,000* in number. They would inevitably perish, if our support was withdrawn from them. The People of this Country, are more inclined to fire upon than receive & protect a Negro whose complaint is the small Pox. The abandoning [of] these unfortunate beings, to disease to famine, & what is worse than either, the resentment of their enraged Masters, I should conceive ought not to be done, if it can possibly be avoided, or in as small degree as the cases will admit.[94]

O'Hara's words touched the earl, but the latter did not want the epidemic raging at Portsmouth to infect his army at Yorktown. "It is shocking to think of the state of the Negroes," Cornwallis confided to O'Hara on 7 August 1781, "but we cannot bring a number of sick & useless ones to this place; some flour must be left for them & some people of the Country appointed to take charge of them to prevent their perishing."[95]

Ten days later, O'Hara added the postscript to this tragic story, which proved to be a foretaste of the tragedy that would engulf a much larger number of runway slaves at Yorktown.

> We shall be obliged to leave over *400* Wretched Negroes. I have passed them all over to the Norfolk side [of the Elizabeth River], which is the most friendly Quarter in our Neighbourhood. I have begg'd of the People of Princess Ann, & Norfolk Countys to take them. We have left with them *fifteen days provisions*, which time will Kill, or Cure the greatest number of them, such as . . . will by that time, be free from the small Pox,—which is the invincible objection, the people have, to these miserable beings.[96]

---

[92] Feltman, *Journal*, p. 6.
[93] Ltr, Leslie to Cornwallis, 13 Jul 1781, P.R.O. 30/11/6, Cornwallis Papers.
[94] Ltr, O'Hara to Cornwallis, 9 Aug 1781, P.R.O. 30/11/70, Cornwallis Papers.
[95] Ltr, Cornwallis to O'Hara, 7 Aug 1781, P.R.O. 30/11/89, Cornwallis Papers.
[96] Ltr, O'Hara to Cornwallis, 17 Aug 1781, P.R.O. 30/11/70, Cornwallis Papers.

For African Americans, the Yorktown campaign was a tragedy. What transpired in Virginia in 1781 was the most notable slave uprising to occur in the United States prior to the Civil War. At the bicentennial observances in 1981, François Mitterrand paid those desperate fugitives an unintended tribute when he said, "Everywhere one finds the same desire for independence, the same need for dignity."[97] The African Americans who flocked to Cornwallis registered their hatred for chattel slavery and their desire for liberty—a desire so great they willingly braved the dangers of war to realize it. And thousands chose death instead of returning to bondage. Wherever freedom is cherished, their struggle and their betrayal should be remembered.

---

[97] *Virginian-Pilot* (Norfolk), 20 Oct 1981.

Learning the Good and the Bad: Canadian Exposure to British Small War Doctrine in South Africa, 1900–1901

Chris Madsen

At the beginning of the last century, Canadian troops went to South Africa as part of a multinational force to fight the Boers, a determined and skillful enemy who resorted to unconventional means to carry on the struggle in the face of British superiority. At the height of the conflict, several hundred thousand imperial and colonial soldiers fought against opposition numbering an estimated ten to twenty thousand at most.[1] After a series of initial setbacks, the top British generals took the offensive with substantial forces marshaled from overseas and occupied principal Boer cities and towns in the Orange Free State and Transvaal, a relentless advance that the Boers proved unable to check conventionally. Formed units of Canadian infantry, artillery, and mounted troops participated fully in these operations under the command and control of British superior officers at brigade and above.[2] Unwilling to concede defeat, the remaining Boers opposing the British organized into smaller, decentralized groups, called commandos, and pursued guerrilla tactics of surprise and ambush by striking at points of weakness. In turn, Canadian troops protected extended lines of communication, conducted patrols to ward off attacks and to gain information on Boer strength and intentions, and eventually joined British efforts to put pressure on Boer fighters by calculated measures against the civilian population. As the changed nature of the conflict was grudgingly acknowledged by higher commanders in the field, corresponding demands on individual soldiers, who for the most part longed to go home, grew tiresome and frequently disagreeable. It was not the type of warfare that either the troops on the ground or the government that had sent them envisioned when the commitment was made.

As the country's first significant overseas deployment in a combat role, the South African War (also known as the Boer War) introduced Canada's nascent armed forces to a complex battle environment that tested accepted

---

[1] Alexander Hugh John Wilson, "The Railway War: A Study and Assessment of the British Defeat of the Boer Guerillas, South Africa 1899–1902" (Master's thesis, University of New Brunswick, 1993).

[2] Brian A. Reid, *Our Little Army in the Field: The Canadians in South Africa 1899–1902* (St. Catherines, Ontario: Vanwell, 1996).

knowledge with practical experience gained in the field against a flexible foe. The Canadians underwent a learning process, in the course of which the distinct stages of instruction, adaptation, and evolution pertained. Officers of the Canadian permanent force were predominantly schooled in British-inspired small war doctrine described in manuals and taught through courses of military instruction. Firsthand experience in the 1885 North-West Rebellion for some, along with recent best practice in Great Britain's colonial wars, particularly on India's northwest frontier, provided important sources of inspiration and dissemination. Field Marshal Lord Roberts (Frederick Sleigh), Baron of Kandahar, and his chief of staff, Maj. Gen. Lord Kitchener (Horatio Herbert), Baron of Khartuom, (the duo sent to South Africa to retrieve British fortunes and deliver victory), imbued the operational style of the Indian Army onto the large field force collected to march on Bloemfontein and onward to Pretoria. As part of this campaign, Canadian tactical units integrated into larger groupings, commanded by superior British officers, which fought the Boers in numerous battles and tactical engagements. Constant losses to disease, combat, and mishap drained manpower and combat capability to the point that diminished Canadian units were largely relegated, once the conflict entered the guerrilla phase, to static duties along railroads and places considered important for maintaining lines of communication. Opportunities for independent command at lower levels and the need to confront the Boers on their own terms enabled the Canadians to develop their own style of fighting, based upon working in small combined arms teams and as part of task-oriented columns organized by the British. Canadian troops reached the pinnacle of combat efficiency and competence just as higher British commanders decided to shift operations toward civilians and the return to Canada became a viable option. Canadians left the hard business of subduing intransigent insurgents and actually winning the conflict to others who were better suited to undertake the necessary work.

Instruction in British Example

At the turn of the twentieth century, the Canadian armed forces were militia-based, with a small cadre of permanent force officers and other ranks for training purposes. Withdrawal of British garrisons in the previous decades put a greater burden for the organization of defense on the Canadians themselves. They had last conducted a major mobilization during the 1885 North-West Rebellion, a domestic aid to a civil power operation characterized more by the long distances of transporting a sizable force of troops from central Canada to the prairie territories and sustaining them in the field rather than by the smallish battles and engagements actually fought to reassert dominion authority.[3] Propitiously,

[3] Desmond Morton, *The Last War Drum* (Toronto: Hakkert, 1972); Walter Hildebrandt, *The Battle of Batoche: British Small Warfare and Entrenched Metis* (Ottawa: National Historic Parks and Sites, Canadian Parks Service, Environment Canada, 1989).

the rebels under arms, led by Louis Riel, chose not to adopt guerrilla tactics against vulnerable supply lines and instead waited for the field force to come to them. The North-West Rebellion gave a select number of officers practical experience in command at levels of responsibility below brigade and limited exposure to combat in a Canadian context.[4] Few important lessons, even the most obvious for dealing with insurgents, carried over to the Canadian militia, which lapsed into a period of apathy and heavy desertions from its ranks. The primary external military threat to Canada at the time was the United States, a prospect considered increasingly untenable due to a totally indefensible border and undue faith in the effectiveness of British sea power on two nations sharing a continent. In 1898, a report by a commission into the state and improvement of Canadian defenses concluded that the country could not possibly defend itself without significant assistance from imperial troops, who likely would arrive neither in time nor in sufficient numbers to prevent a full-scale land invasion.[5] Thus, provision of military forces was framed more in terms of extraterritorial service as a contributing member of the British Empire, particularly after Maj. Gen. Edward Hutton became general officer commanding the Canadian militia. With the outbreak of hostilities in South Africa, this British officer made immediate plans for sending a large Canadian contingent drawn from the militia. However, when the public learned of this, it put Canada's prime minister, Wilfrid Laurier, and his defense minister, Frederick Borden, into a very awkward situation politically.[6] Fervor for the war was strongest among English-speaking parts of Canada, whereas opinion in French-speaking, Catholic Quebec was decidedly against participation in any imperial war that suppressed another religious and ethnic minority group. The Canadian government compromised by indicating a willingness to send contingents of volunteers for a fixed period of one-year service, to be placed under British imperial authorities in the conduct of operations. The British were left with little choice in the matter other than to accept this restriction by a self-governing colony.

British approaches to training and doctrine underpinned recruitment of contingents destined for South Africa from Canada. Though the newest recruits among other ranks came from various walks of life and often possessed little or no direct military experience, permanent force officers and noncommissioned officers were overwhelmingly represented in positions of authority across battalions, squadrons, and batteries. These men were familiar with the latest

---

[4] Maj A. N. Todd Diary, Mss C550/2/2.1, University of Saskatchewan Library Special Collections.

[5] Memo, Maj Gen E. P. Leach for Under Secretary of State for War, 30 Nov 1898, WO 32/6366, The National Archives (TNA), Kew, United Kingdom. See also Stephen J. Harris, *Canadian Brass: The Making of a Professional Army, 1860–1939* (Toronto: University of Toronto Press, 1988), pp. 62–64.

[6] Carman Miller, *Painting the Map Red: Canada and the South African War, 1899–1902* (Montreal and Kingston: McGill-Queen's University Press, 1993), ch. 3; Desmond Morton, *Ministers and Generals: Politics and the Canadian Militia, 1868–1904* (Toronto: University of Toronto Press, 1970), pp. 151–56.

manuals issued by the War Office, inspected annual summer camps of exercises performed by the active militia, and staffed established schools of instruction. The focal point of military professionalism in Canada was Toronto, with its long-standing infantry school at Stanley Barracks and the private Canadian Military Institute (later the Royal Canadian Military Institute), which sponsored regular lectures and served as a meeting place for the exchange of views.[7] Lt. Col. William Otter, the infantry school's commandant, was designated to take a special service battalion of the Royal Canadian Regiment (RCR) to South Africa, first saw active service during the Fenian Raids, commanded the Battleford column during the North-West Rebellion, and wrote a popular book published in several editions about administration of an infantry battalion. Lt. Col. Francois Lessard, a French-Canadian officer also associated with the military district in Toronto, was a cavalry proponent earmarked to lead mounted troops in the second South African contingent.[8] Reformist British officers like Hutton, who fell afoul of the traditional cavalry lobby in the British Army, imported new concepts to Canada about the relative value of mounted infantry. This built upon a previous book by George Taylor Denison, a lawyer and part-time soldier from a well-known Toronto military family, titled *Modern Cavalry: Its Organization, Armament and Employment in War*, which won a prize for military writing from the Russian czar.[9] In western Canada, the North-West Mounted Police provided a ready source of capable riders, and Lord Strathcona's Horse, a dedicated unit of mounted infantry, was formed through the generosity of Great Britain's high commissioner in Ottawa, who paid the associated costs. The various formations adopted standard British organization as patterned at the time and set out in existing manuals. On the theoretical side, the Canadians were at least comparable to similar military units in Great Britain and other self-governing colonies in the British Empire.

The accepted knowledge that formed prevailing sources of British doctrine on small wars was available to Canadian officers if they chose to study it. A second edition of Maj. Charles Callwell's *Small Wars* was published in 1899 and already used as a teaching text in such places as the staff college at Camberley.[10] It became available under official auspices for purchase through Her Majesty's Stationery Office, conveniently by mail order for addition to military libraries or personal collections. Callwell highlighted the peculiar characteristics of fighting against irregular enemies and drew upon

---

[7] Col. Gerald Charles Kitson, "Lessons in Strategy and Tactics to Be Gained from the Present Campaign" (lecture delivered on 12 March 1900), no. 10, *Canadian Military Institute Selected Papers* (Toronto: Canadian Military Institute, 1900), pp. 22–40.

[8] John Macfarlane, "The Right Stuff? Evaluating the Performance of Lieutenant-Colonel F. L. Lessard in South Africa and His Failure to Receive a Senior Command Position with the C.E.F. in 1914," *Canadian Military History* 8 (Summer 1999): 50–51.

[9] George Taylor Denison, *Modern Cavalry: Its Organisation, Armament and Employment in War*, CIHM 02636 (London: T. Bosworth, 1868).

[10] For curriculum and teaching at Camberley during this period, see Brian Bond, *The Victorian Army and the Staff College, 1854–1914* (London: Eyre Methuen, 1972).

illustrations from recent operations by the British and other armies in the colonial sphere, observing, "Tactics favour the regular army while strategy favours the enemy—therefore the object is to fight, not to maneuver."[11] By virtue of technical and professional superiority, conventional forces generally bested opposing colonial enemies in open battles and tactical engagements but responded awkwardly to actions done in a less forthright manner, which professional soldiers were apt to describe as underhanded or cowardly. For a British Army at home aping Prussian forms and still entranced by the marvelous conventional victories of Helmuth von Moltke in the wars of German unification (the handling of *francs tireurs* after the French Army surrendered at Sedan was typically overlooked), Callwell's book and the peculiar demands of colonial warfare probably received less attention than they deserved.[12] Nonetheless, each year, a number of Canadian graduates from the Royal Military College of Canada obtained commissions in the British Army and served with regular regiments throughout parts of the British Empire. Henry De Bury, one such officer who soldiered in Southwest Africa, wrote Canada's first significant counterinsurgency treatise, *Bush Wars*, a blend of simple theoretical analysis and personal reminiscences.[13] Private reading and writing supplemented formal courses of instruction at Canadian military educational institutions. The Royal Military College of Canada ran the first higher-staff course, modeled by Col. Gerald Kitson upon Camberley's curriculum, between 1 February and 27 May 1899; eight of the twelve officers who completed the course proceeded to active service in South Africa.[14] A tension inevitably exists between what is taught in an abstract way and what soldiers actually encounter in the field through practical experience. Hopefully, Canadian officers went to South Africa better prepared than they would otherwise have been without reading manuals and undertaking instruction. But, in the end, the operational commander, guided by his background, personality, and views, decided the concept of operations followed in the specific campaign.

Field Marshal Lord Frederick Roberts, once he assumed command in South Africa on 10 January 1900, planned to undertake offensive operations against the Boers. His ideas were influenced by his time in India and followed classic textbook tenets. The Indian Army had faced possible invasion from Russia through Afghanistan as well as pacification of various tribal peoples on the northwest frontier, to which Roberts

[11] Charles Edward Callwell, *Small Wars: A Tactical Textbook for Imperial Soldiers* (London: Greenhill, 1990), ch. 7.

[12] Howard Bailes, "Patterns of Thought in the Late Victorian Army," *Journal of Strategic Studies* 4 (1981): 29–45.

[13] Thanks to Maj. Andrew Godefroy for bringing this text to my attention.

[14] "Department of Militia and Defence for the Dominion of Canada. Report for the Year Ending December 1899," *Sessional Papers* 34, no. 19 (1900): 33. See also John A. Macdonald, "In Search of Veritable: Training the Canadian Army Staff Officer, 1899 to 1945" (Master's thesis, Royal Military College of Canada, 1992), pp. 62–63.

espoused a "scientific frontier doctrine" of presence and forward defense.[15] The situation in South Africa was simply the reverse side of the problem, wherein the British were determined to take the initiative away from the Boers. Roberts intentionally stayed on a defensive footing for the time being: "The conclusion that I arrived at was that no sensible improvement to the military situation could be hoped for until we were prepared to carry the war into the enemy's country, and all my efforts have accordingly been exerted in that direction."[16] With fresh reinforcements arriving from overseas, upwards of 35,000 troops with supporting cavalry and artillery were moved forward and concentrated in sufficient force for a planned general advance on the Orange Free State and then onward to Transvaal, thereby drawing off the Boers and relieving besieged British garrisons elsewhere.[17] The first Canadians in South Africa, the 1,000-strong infantry battalion under Colonel Otter, received orders to join the 19th Brigade under Maj. Gen. Horace Smith-Dorrien, belonging to Lt. Gen. Henry Colville's 9th Division. Brev. Maj. Septimus Denison, an RCR officer from the same Toronto family, joined Roberts on his staff as Canadian aide-de-camp. The intended line of advance was eastward across the open countryside along the Modder River, a movement by which Roberts intended to cut loose from the main railway line and thus surprise the Boers as to his intentions. In due course, the British cut off and cornered a large force of five thousand Boers retreating from Kimberley to Bloemfontein, commanded by General Piet Cronjé, at Paardeberg Drift, where a major battle ensued. The Canadians took heavy casualties during a frontal assault ordered by Kitchener on the first day and eventually were instrumental in a night action prepared by Otter and Smith-Dorrien that led to Cronjé's surrender.[18] In keeping with Callwell's writings, Roberts gave the trapped Boers little option other than to fight superior forces arrayed against them, in which British materiel and numbers prevailed. In the weeks and months ahead, the Boers proved powerless to stop by conventional means the plodding advance of Roberts' massive field army toward Boer cities, the capture of which Roberts felt would compel surrender and put an end to the war.

---

[15] R. A. Johnson, "Russians at the Gates of India? Planning the Defence of India, 1885–1900," *Journal of Military History* 67 (2003): 697–744; Tim Moreman, *The Army in India and the Development of Frontier Warfare, 1849–1947* (Basingstoke: Macmillan, 1998).

[16] Rpt of Operations, Lord Roberts to Secretary of State for War, 6 Feb 1900, WO 32/7962, TNA.

[17] Circular Memo 5, Notes for Guidance in South African Warfare, 26 Jan 1900, WO 108/109, TNA.

[18] Rpt, Smith-Dorrien to Assistant Adjutant General, 9th Division, 1 Mar 1900, box 87/47/6, Boer War Notes binder, General Horace Smith-Dorrien Papers, Imperial War Museum (IWM), United Kingdom; Weekly Rpts, Otter to Chief Staff Officer, Ottawa, 26 Feb and 2 Mar 1900, RG 9 series II-A-3 reel T-10404 vol. 32, Library and Archives Canada (LAC), Ottawa.

Adaptation to Unconventional Methods of Warfare

British forces entered Bloemfontein, the Orange Free State's capital, on 13 March 1900 without significant resistance. Boer commanders chose to withdraw instead of meeting British strength head-on in battles they knew were losing propositions. The refusal of the Boers to accept the fight on British terms created a dilemma for Roberts and his plans.[19] Boer fighting forces remained mostly intact, and various attempts to draw them into action failed. In fact, the Boers nearly handed the British a number of defeats by taking advantage of British mistakes at the tactical level. Sent to the rescue of Brig. Gen. Robert Broadwood's ambushed column on 31 March, Canadian infantry played an ancillary part in an engagement described by Smith-Dorrien as "a very hard day for the men."[20] The slow-moving infantry was hardly a match for the mobility of the Boers on horseback. Mounted infantry, rather than the ponderous cavalry favored by senior British officers of that branch at the time, were the answer, but still too few in number and seldom used to full potential. To illustrate, British commanders broke the mounted infantry and artillery in the second Canadian contingent into piecemeal groupings and distributed them on various tasks. However, Lessard's 1st Canadian Mounted Rifles found a home with Hutton's brigade in Lt. Gen. John French's Cavalry Division.

When the operational pause forced on Roberts by logistics concerns ended, the advance resumed at the end of April in accordance with the previously laid planning. The 19th Brigade with the RCR infantry transferred from the control of 9th Division to a new force or column organized under Lt. Gen. Ian Hamilton at Winburg. On 2 May, Smith-Dorrien assumed command of the infantry division in Hamilton's force, of which the 19th Brigade formed a part. Canadian confidence in British leadership generally remained high, though the same was not always true in the other direction. Hutton criticized Lessard's tactical competence, and Lt. Col. Lawrence Buchan took over command of the RCR for a period of time when Otter suffered a superficial injury in battle and went back to hospital.[21] The last conventional battles awaiting the Canadians on the road to Pretoria were among the hardest faced in South Africa.

In spite of several concerted Boer attempts to block the advance, the British and Canadians worked as a seamlessly integrated team to achieve objectives quickly and limit casualties. The shared experiences of the previous weeks and the common operating doctrine paid handsome dividends in terms of tactical success. Looking for another Paardeberg through a decisive battle with large Boer forces, Roberts was optimistic when the

---

[19] Maj. Andrew J. Risio, "Building the Old Contemptibles: British Military Transformation and Tactical Development from the Boer War to the Great War, 1899–1914" (Master's thesis, U.S. Command and General Staff College, 2005), p. 3.

[20] Diary, 31 Mar 1900, box 87/47/1, Smith-Dorrien Papers, IWM.

[21] Serge Derflinger, "Otter's Wound and Other Matters: The 'Debate' Between William Dillon Otter and Lawrence Buchan," *Canadian Military History* 7 (Autumn 1998): 60.

Boers stopped behind the Zand River. On 10 May, Smith-Dorrien's two infantry brigades attacked across the water obstacle, while the Canadians fixed approximately eight hundred Boers along the riverbank by rifle and artillery fire.[22] The Canadians lost only one killed and several wounded for a full day's fighting. Though Robert turned the Boer position shortly thereafter, several thousand Boers managed to get away on horseback. The foot-borne 19th Brigade gave pursuit as best as possible; at one point, Smith-Dorrien's infantry caught up with 11,000 Boers, field guns, and wagons, only to see the whole mass promptly ride away before their eyes.[23] The Boers, for the most part, avoided armed confrontations, unless certain natural geographic features gave a clear and overwhelming advantage. Outside Johannesburg, the Boers fortified a series of high kopjes around Doornkop east of the Klip River and installed heavy artillery, which fired in a commanding fashion onto the plains below, blocking the way forward. On 29 May, Smith-Dorrien directed the Canadians and Scots in the Gordon Highlanders Regiment to advance up a steep, grassy slope, which the Boers promptly set alight, and to undertake an assault.[24] The action went much better than expected, due in large part to the coordinated efforts of the units involved. In all, the Canadians lost only seven wounded.

Roberts arrived in Pretoria, which Boer military forces abandoned, with considerable fanfare on 5 June 1900. Smith-Dorrien issued a brigade order to mark the occasion.

> The 19th Brigade has achieved a record of which any infantry might be proud. Since the date it was formed, 12 Fe[bruary 19]00, it has marched 620 miles often on ½ rations seldom on full. It has taken part in the capture of ten towns, fought in ten general actions, and on 27 other days. In one period of 30 days it fought on 21 of them and marched 327 miles. Casualties [were] between 4 & 5 hundred.[25]

For the Canadians, the intensity of combat and range of engagements were unprecedented: 27 officers and 411 other ranks remained effective in the battle-weary RCR, less than half full strength. It seemed an opportune time to go home, as Roberts and the soldiers alike believed they had finished the war. British professional soldiers immersed in existing doctrine sincerely held that occupation of capital cities was key to victory by putting the enemy into an unfavorable—indeed, potentially unwinnable—position. Many Boers, civilians

[22] Rpt, Buchan to Chief Staff Officer, Ottawa, 31 May 1900, RG 9 series II-A-3 reel T-10404 vol. 32, LAC.

[23] Diary, 20 May 1900, box 87/47/1, Smith-Dorrien Papers, IWM.

[24] Rpt, Smith-Dorrien to Assistant Adjutant Gen Hamilton's Force, 31 May 1900, WO 105/8, TNA.

[25] Notebook, Pvt James Montgomery Thomas, 19820205-004, Canadian War Museum (CWM), Ottawa.

under arms fighting for their homeland, refused to admit defeat and embarked on guerrilla attacks against British weak points and lines of communication.

The shift of the conflict into an irregular phase invited a slow response from Roberts and the British. After Pretoria's surrender, large numbers of armed Boers still loitered in the general vicinity. Pleas to turn themselves in went unanswered, and every attempt to engage them with the tired forces available proved fruitless, no matter how hard Roberts and his subordinate commanders tried. Mopping-up operations gave way to dealing with numerous attacks on railways and telegraphs along lines of communication between occupied cities and towns. Even though vastly superior in numbers, the British field army could not be strong everywhere over a wide geographical area. The Boers split up into smaller organized commandos directed locally in a decentralized fashion. The intransigent enemy general, Christiaan De Wet, led by example and encouraged his followers to tear up tracks, blow up trains, cut wires, and hit the British where most vulnerable.[26] Roberts initially considered such attacks a mere nuisance and made provision for the better protection of assets deemed valuable. He distributed the 19th Brigade, supported by mounted infantry and artillery, along the railway line between Pretoria and Kroonstad, and placed Smith-Dorrien in command over lines of communication on 11 June. The Canadians moved to Springs, a coal-mining town located on a branch line off the main railway, and established a garrison. Dispersal of British strength in this way contravened basic military principles, though the decision was understandable given thinking at the top and the perceived nature of the threat. Roberts wrongly assessed that the Boers lacked the capacity to carry on the fight for an extended period of time.[27] The policy of burning farms, subsequently introduced in retribution to persistent Boer attacks, proved a further misstep and a significant escalation of matters.

An attack on a construction train repairing damaged track on 14 June 1900 provided the immediate justification for implementation of farm burning. Based on this incident, Roberts ordered Kitchener to have De Wet's farm burned down and issued the following instructions to Smith-Dorrien: "You should let it be known that if the Railway or telegraph lines are damaged the nearest farm to the break will be burnt to the ground and in the event of such damage occurring in your section of the line you should at once burn the nearest farm. This will I think have a salutary effect."[28] Besides the obvious moral and legal complications of applying this order indiscriminately toward civilians not directly involved in attacks or raids, the promulgated policy rested on questionable assumptions from an operational perspective, the least of which was that Boer combatants would stop under such pressure. In reality, farm burning

[26] Christiaan De Wet, *Three Years' War* (New York: Scribner, 1902).

[27] MS, Maj Rob B. McClary, Learning the Hard Way, or Not at All: British Tactical and Strategic Adaptation During the Boer War, 1899–1902 (School of Advanced Military Studies, U.S. Army Command and General Staff College, 1999), pp. 23–24.

[28] Telg C.2087, Roberts to Smith-Dorrien, 14 Jun 1900, WO 105/36, TNA.

dispossessed more people and just gave the Boers greater reasons to keep on fighting.[29] Roberts sought a military solution to a narrow problem but, instead, produced an unintended effect that impeded reaching strategic and operational objectives in South Africa. Troops in the field, like the Canadians, carried out the task of turning out civilians and burning down farms reluctantly and only because they were ordered to. Smith-Dorrien commended several officers of the 2d Canadian Mounted Rifles for their farm-burning work wherever a breach or damage occurred to the railway. Lessard's 1st Canadian Mounted Rifles and Lord Strathcona's Horse, meanwhile, joined in active operations chasing the elusive Boer commandos. A Canadian lieutenant serving in the Brabant's Horse wrote home: "The war is getting very tiresome. De Wet is still at large and we are practically doing police work."[30] Heretofore, the British and Canadians had been reactive in adapting to changing Boer tactics. Senior leaders gradually disassembled the larger organizational groupings suitable for conventional warfare and gave greater discretion to lower commanding officers and subordinates, who exercised opportunities for independent decision and action. Out of this field experience emerged a distinct Canadian approach to fighting the Boers on their own terms.

Evolution Toward a Canadian Style of Small Warfare

Operations against the Boers necessitated reevaluation of the existing doctrine applied by the Canadians and the relationship with immediate British superiors. Canada never really had an effective reinforcement policy to replace losses in South Africa, and the effects of disease and combat casualties lowered personnel strength. Months of hard fighting and field operations through the spring and summer reduced infantry and mounted rifle units to less than half strength, with proportionately diminished combat capability. The field artillery was little better off, due to widely distributed deployment. Canadians accustomed themselves to making do with fewer persons and inadequate equipment. For example, most of the horses brought over from Canada in the second and third contingents perished due to poor diet and exhaustion brought on by overwork; they were replaced by inferior grades of horses and, in many cases, ponies. Scattered Canadian troops were brought together once again to perform duties along lines of communication in the Transvaal farther up the main railway line past Pretoria, pending arrangements to move back to Cape Town for return to Canada. Units were generally too weak for much other useful employment; having the Canadians together under Canadian officers at least provided some critical mass. The 1st Canadian Mounted Rifles, now known as the Royal Canadian Dragoons (RCD), garrisoned at Belfast, a town captured

---

[29] Chris Madsen, "Canadian Troops and Farm Burning in the South African War," *Canadian Military Journal* 6 (Summer 2005): 55.

[30] Ltr, Ramsay to sisters, 24 Sep 1900, R3177-0-8-E, Frederick Ramsay, LAC.

on 24 August by the 11th Division roughly halfway to the railway's terminus at Komati Poort. The British intended to interdict Boer supplies coming through Portuguese territory and to establish a firm British presence in the area leading to Transvaal's full annexation. Taking advantage of challenging geographical features in the mountainous and hilly countryside, the Boers still roamed freely and resisted British incursions in a strong manner. A section of Royal Canadian Field Artillery under the command of Lt. Edward Morrison from Battery D also joined the Belfast garrison. Lessard, who returned from sick leave on 27 August, technically reported to Hutton for the time being; while in actuality, he answered directly to Lt. Col. J. W. Godfray, the station commander. This defensively minded British officer prohibited the Canadians from straying too far from Belfast, for fear of either having to come to their rescue or inviting a Boer attack on the town.

The RCD manned advance outposts on the edge of Belfast and performed limited patrolling until Lessard's appointment as acting station commander in the first two weeks of October allowed for stepped-up activity. On 5 October 1900, the Canadians conducted a reconnaissance in force consisting of sixty mounted troops and Morrison's field guns as far as Weltevreden and fought a small action with the Boers.[31] The operation, planned and executed by Lessard at his initiative, gained information on Boer intentions and signaled the start of aggressive patrols from Belfast into the hinterland. The Canadian officer reasoned that it was foolish to wait for the Boers to make the first move; he instead sought to unsettle the enemy as much as possible through surprise and targeted action when the opportunity arose. RCD troopers appeared where least expected and became better aware of local conditions. The change from a defensive to a more offensive stance anticipated Kitchener's selection of Smith-Dorrien to organize a flying column based out of Belfast.

Preparation and organization of forces marshaled at Belfast reflected subtle changes in the mode of operating against Boer irregulars. South of the town, the Carolina and Ermelo commandos, acting together, mauled British forces covering the pullback of French's cavalry and afterward regularly cut the railroad.[32] The senior leadership responded by forming numerous flying columns to seek out meddlesome Boer armed groups and destroy the basis for their support, in the way of foodstuffs and living dwellings. The descriptive word "flying" was somewhat misleading, since the predominantly infantry composition and ponderous supply requirements of the columns limited movement compared to the mobility of the faster and more flexible Boers. Alas, British columns could not stay away from a home base too long and lacked speed and size when confronted with a local enemy superior in numbers. During a visit to Pretoria on 26 October to watch Transvaal's annexation

[31] RCD War Diary, 5 Oct 1900, RG 9 series II-3-A reel T-10404 vol. 32, LAC.

[32] Rpt of Operations, Roberts to Secretary of State for War, 3 Jan 1901, WO 32/8001, TNA. French's cavalry was withdrawing because the climate in the area was making the British horses sick.

parade and a troop review, Kitchener told Smith-Dorrien of his assignment to command the column operating from Belfast, augmented by three infantry battalions and mounted troops, "to assume active operations against [the] enemy."[33] Lessard's Canadians and the Royal Irish Regiment were already there and familiar with the local situation. By this time, Kitchener was the real driving force because Roberts, appointed to be the next commander-in-chief of the British Army, was preparing to depart for London. Hardly bothered by humanitarian scruples, Kitchener intended to take the war directly to the Boer populace in an effort to wear down and outlast Boer fighters. Irregular opponents demanded unconventional methods with which the Canadians were already acquainted. The 2d Canadian Mounted Rifles, now known simply as the Canadian Mounted Rifles (CMR), sent a detachment of three troops to Belfast to join Lessard and Morrison under Smith-Dorrien's direct command and control. Next, Smith-Dorrien finalized preparations and plans for several major expeditions in early November with Lt. Gen. Neville Lyttelton, the general officer commanding the 4th Division through whom Smith-Dorrien reported.

Operationally, the initial excursion proved a huge disappointment. Battalions drew three days' rations for troops and horses ready for movement at short notice. The intent from the beginning was to get out and back quickly in a day or two. On the morning of 1 November, Smith-Dorrien briefed commanding officers on his plan to advance two separate columns southward independently, rendezvous at a designated location, and then proceed to attack a main camp and several houses believed to be used by Boer commandos for comfort and supply near Witkloof on the Komati River. Lessard's RCD squadron and Morrison's field guns provided the advance guard for the right column led by Lt. Col. James Spens of the Shropshire Light Infantry, while the CMR accompanied the left column led by Smith-Dorrien. Whereas splitting the force from the outset for no apparent good reason constituted an unconscionable error, and departure deferred until the early evening potentially compromised operational security. The weather worsened, with thunder, hail, driving rain, and falling temperatures, but instead of turning back, the troops had a very miserable march in the dark and spent a cold night with little sleep out in the open. Early next morning, the CMR became separated from the main column and engaged some Boers, during which Lt. T. W. Chalmers was killed trying to retrieve another wounded officer. By now, Smith-Dorrien realized that he had lost the element of surprise, and the bedraggled state of his troops probably made carrying on as planned a risky venture.[34] He ordered the supply transport and columns back to Belfast, while a rear guard kept the Boers at a respectable distance during the withdrawal. His countrymen accorded Chalmers a funeral

[33] Diary, 26 Oct 1900, box 87/47/1, Smith-Dorrien Papers, IWM.

[34] Rpt, Smith-Dorrien to Chief Staff Officer, 4th Division, Middleburg, 4 Nov 1900, WO 105/12, TNA.

with full military honors and mention to Roberts for his bravery in an otherwise abortive operation.[35] For the Canadians, the mission was hardly a testament to British tactical leadership, which would be tested again during the next expedition that resulted in the battle that Canadians know as Liliefontein.

Liliefontein, among Canada's most celebrated feats of arms in South Africa, demonstrated the extent to which Canadian troops worked in small combined arms teams in the face of a determined enemy. Smith-Dorrien departed early on 6 November with the CMR attached to a main column and Lessard leading the RCD and Morrison's field artillery with the advance troops. The mission picked up from where the previous expedition had left off, with the objective of destroying houses of known or potential use to the Boers along the Komati River.[36] The Boer commandos, alerted to the advancing British, made a stand near Witkloof due to a large Boer supply convoy crossing the waterway; in other words, the British and Canadians threatened something really important to them worth defending. The infantry and field artillery fought all day and finally forced the Boers to retire back across the river, leaving Smith-Dorrien's force to camp on high ground near Liliefontein that night. Instead of crossing the river the next morning, Smith-Dorrien decided to return to Belfast because surprise again had been lost, and the Boers were being reinforced. (Unbeknownst to the British, Boer commanders planned their own attack on the Liliefontein encampment.)

The CMR beat the Boers in a mad gallop to the heights commanding the route back, as Lessard, the RCD, Morrison's field guns, and a Canadian-manned Colt gun covered the rear of Smith-Dorrien's departing column. The Boers made repeated attempts over several hours to capture the guns as the mounted troops fought them off. Morrison described an eventful part of the action:

> The Boers were certainly coming on with determination at that point. I went into action [with the guns] and soon scattered the mounted men and they dismounted and came on running from cover to cover and my gunners were soon exposed to a sharp rifle fire. Lieut Cochburn sent his men further to the front and we were getting this rush under control when Col. Lessard galloped up and said: "For God's sake, Morrison save your guns! They are coming down on our flank." He pointed out to the left (I speak always with relation to the front of the column—not of the rear guard) and as I looked the Boers

[35] Telg Z.390, Smith-Dorrien to Roberts, 3 Nov 1900, WO 105/15, TNA. Maj. Richard Turner of the RCD recorded, "Have just returned from the worst 24 hours I have experienced in Africa. We left camp at 5.30 pm marched until 11 pm raining heavily, and bitterly cold. From then until 4.30 am lined to 'stand to' no fires–no smoking expecting attack. Trekked at daylight and quite a bit of fighting all day. On the way back the gun limbers were piled up with Gordons played out from exposure. . . . Almost half our men are now down with rheumatism." Diary, 4 Nov 1900, 19710147-001, General Sir Richard Ernest William Turner, CWM.

[36] Hugh John Robertson, "The Royal Canadian Dragoons and the Anglo-Boer War, 1900" (Master's thesis, University of Ottawa, 1983), p. 182.

> were coming on for half a mile on our flank to cut us off from the ridge above the spruit which was to be our next position on retiring.[37]

Small parties of Canadians blunted the final Boer charge long enough for the guns to reach safety. The Boers overran and captured sixteen troopers, besides inflicting fourteen casualties killed and wounded, including several officers. But the Canadians under Lessard and Morrison had held off a superior force with considerable competence and coolness under fire. In recognition of individual acts of bravery, Canadians received the highest British military honors: three Victoria Crosses and a Distinguished Service Order. The achievement of the Canadians overshadowed the fact that the Boers had chased Smith-Dorrien and his column back to Belfast. A different sort of expedition involving the Canadians before they left for Canada met markedly less resistance in the quiet area north of Belfast.

Encouraged by Kitchener, Smith-Dorrien designed the next operation leaving on 13 November as a punitive one to put pressure on the civilian population. Refugee women and children were turned out of Belfast onto the open veldt as the first step of psychological intimidation to say that the British were coming. A column comprising 80 RCD, 2 Colt teams, and Morrison's 2 field guns among the advance troops and 60 CMR in the main body drove through Boer outposts and proceeded into hostile territory, cutting a swath of destruction in its path to sow despair among the Boers and convince them to stop fighting. The expedition, which lasted the better part of four days, worked in conjunction with another column from Middleburg, moving up either side of the Steelpoort Valley, destroying and plundering in tandem.[38] Troops herded the elderly, women, and children outside to watch their homes and belongings go up in flames, the livestock carried away, and the crops destroyed. A Canadian bombardier described the effect on the once picturesque community:

> a fertile valley well populated burning the houses and ravaging the mills completely devastating the place ruining the owners leaving slight shelter for the women and children. . . . Having searched and burned Whitpoort, we started back leading to the right marching our route by burned and ruined homesteads. . . . The ravages of war began. The beautiful street lined with trees and foliage spring water ponds and shady nooks covered with debris from the dynamited houses all the dwellings burned and looted nothing being left for the helpless populace except a Dutch church. All the adjoining farms were ransacked and burned.[39]

---

[37] Rpt, Morrison to Officer Commanding, Royal Canadian Field Artillery, 15 Nov 1900, RG 9 series II-A-3 reel T-10404 vol. 33, LAC.

[38] Rpt, Smith-Dorrien to Chief Staff Officer, 4th Division, Middleburg, 18 Nov 1900, WO 105/12, TNA.

[39] Diary, 14–16 Nov 1900, 19940001-831, CWM.

Canadian involvement in harsh and ruthless British methods against civilians was a distinguishing feature of evolving doctrine of how to deal with irregular opponents who refused to give up in South Africa. They accepted the work only grudgingly, as Morrison wrote, "It was a terrible thing to see, and I don't know that I want to see another trip of the sort, but we could not help approving the policy, though it rather revolted most of us to be the instruments."[40] This last employment around Belfast provided a brief introduction to the nastier aspects of counterinsurgency operations, one the Canadians would not have to repeat as the RCD, CMR, and Canadian field artillery readied to leave South Africa, to be followed by Lord Strathcona's Horse the following month. Once Kitchener replaced Roberts as operational commander, the fight against insurgents and policies toward civilians became systematic. Instructions issued in December 1900 urged,

> Officers commanding columns that they should fully recognize the necessity of denuding the country of supplies and livestock, in order to secure the two-fold advantage of depriving the enemy the means of subsistence, and of being to feed their own columns to the fullest extent from the country. These, and not the destruction of farms and property, should be the objects of all columns, second only to the actual defeat of the enemy in the field.[41]

Eradication of crops and similar forms of sustenance took top priority. Irregular conflict carried on for another two years, while Canada was called upon to send more troops.

The departure of the Canadians evoked mixed emotions among higher British military commanders. Smith-Dorrien, their immediate superior officer, never begrudged soldiers who had fought for months and made a measurable contribution for wanting to return home. He thanked the RCD

> for the grand work they have performed for him in the Belfast Flying Column. In 8 of the last 19 days they have been engaged with the Boers and have proved themselves splendidly brave and mobile mounted troops. . . . He can merely say that he would choose no other Mounted Troops in the world before them if he had his choice and he sincerely hopes the day may come when he may have them again under his command.[42]

Withdrawal of almost all available mounted troops at Belfast restricted operations and halted any further expeditions until more arrived. Insufficient troops for the tasks required remained the key constraint in British operations. Roberts, who also departed in late November, saw the Canadians leave "with

---

[40] E. W. B. Morrison, *With the Guns in South Africa* (Hamilton: Spectator Printing, 1901), pp. 277–78.

[41] Circular Memo 27, 7 Dec 1900, WO 108/109, TNA.

[42] Special Orders, Maj Gen H. L. Smith-Dorrien, D.S.O. Commanding Pan, to Dalmanutha, 20 Nov 1900, 19730069-001, Turner, CWM.

deep regret, not only on account of their many soldierly qualities but because it materially impaired the mobility and efficiency of the Army in South Africa for the time being, a very critical time, too."[43] In his mind, it only served to prolong the war and give the Boers an opportunity to recover strength. Whatever the politics involved, Canadian soldiers had proven to be good combat troops, suited for the type of warfare in South Africa. An officer on Roberts' headquarters staff held a high opinion of the Canadians: "I never saw finer bodies of men than the Canadian Mounted Rifles and Artillery, and their infantry battalion was one of the best in the army and did splendid work."[44] Admittedly, the Canadians left on a high note with their reputation intact, unmarred by the demoralizing influence of protracted counterinsurgency operations on conventional forces. It was a good time to bring the troops home. On this occasion, Prime Minister Laurier did not ask those soldiers sent to South Africa with a different kind of war in mind at the outset to continue sacrificing their lives and moral integrity for an irregular fight that lacked public support and a clear end. The Canadian government reserved the right to send further troops if requested, especially as the conflict appeared to be a long one.

Conclusion

Canadian involvement in South Africa and exposure to British small war doctrine at the turn of the last century remains instructive for operations against irregular opponents. Canadian troops tactically integrated into broader multinational forces in the field under a foreign operational commander and higher immediate superiors at the division and brigade levels. Common doctrine and organization, disseminated in printed manuals and taught at military schools of instruction, furnished the foundation for collaboration, reinforced and modified by practical experience in the field. The South African War was essentially the first time Canadian soldiers had seen major and sustained combat. They proved adaptable to the tactics of, first, conventional battles and, then, the peculiar demands of countering guerrilla warfare. The Canadians learned the value of preserving the national identity of formed units and the importance of independent command, since foreign senior officers, no matter how well-meaning, could not be trusted to put Canadian interests first. The inflexibility of the British leadership and poor assumptions in the conduct of the campaign were evident, especially as the conflict entered its irregular phase and the Boers adopted unconventional methods. British generals carried the Canadians along on a harsh and ruthless line of action that eventually made insurgents indistinguishable from the civilian population. Due to the effect of losses on small numbers, Canadian units were a shrinking asset in

---

[43] Rpt of Operations, Roberts to Secretary of State for War, 15 Nov 1900, WO 32/8001, TNA.

[44] Notes by Colonel Grierson [n.d.], WO 108/184, TNA.

terms of manpower and combat capability. Since reconstitution was difficult, Canadians made do with the effective numbers available until defined periods of deployment ended and fresh troops arrived. There is never an opportune time to withdraw prior to the end of a conflict without calls of letting down the mission. However, the Canadians, who had become reliable and competent fighters, had performed admirably in months of hard operations and combat, for which many believed it was their turn to go home. Canadians were too nice to consent to the cruel efforts demanded of an operational commander like Kitchener, who eventually delivered a hollow victory at enormous cost in lives and goodwill. South Africa was just not sufficiently important enough for Canada to give up the lives of its soldiers over many years.

Blindness and Contingencies: Italian Failure in Ethiopia (1936–1940)[1]

Richard Carrier

The aim of this paper is to explain why Fascist Italy was incapable of waging a successful campaign of pacification in Ethiopia between 1936 and 1940. The Fascist regime was capable of planning, preparing, and fighting a massive "national war" against Ethiopia in 1935–1936.[2] The capture of Addis Ababa in May 1936 was followed by a long, difficult, often brutal, and inconclusive campaign of pacification. In the end, Mussolini's blindness and at least three contingencies made the pacification of the country an almost impossible task.

Fascist Campaigns in Africa: A Brief Overview

The presence of Fascist Italy on African soil was the result of a mixed desire for prestige, glory, and some sort of social rejuvenation. Mussolini's interest in this continent was to result in a long campaign in Libya that was a real success in an irregular war. In Ethiopia, after six months of uneasy operations and a victory claimed with the capture of Addis Ababa, a pacification campaign began but finally turned into a stalemate. Simply put, "as ventures in applied military force, they differed dramatically from one another in outcome: from Rome's point of view, Libya was a success and Abyssinia a failure."[3]

The Fascist campaign in Libya was a legacy of liberal Italy's attempt to establish itself as a colonial power. It was the continuation of the war orchestrated in 1911 by Prime Minister Giovanni Giolitti against the Ottoman Empire.[4] The campaign that took place in the twenties and the early thirties was a typical colonial military operation: it was a long, frustrating fight against a dedicated and skillful enemy, kept quiet by the government because the

---

[1] Four very close meanings of *contingency:* "The condition of being liable to happen or not in the future," "uncertainty of occurrence or incidence," "the befalling or occurrence of anything without preordination," and "the condition of being free from predetermining necessity in regard to existence or action." *Oxford English Dictionary*, 2d ed., vol. III (Oxford: Clarendon Press, 1989).

[2] Giorgio Rochat, *Militari e politici nella preparazione della campagna d'Etiopia*, ed. Franco Angeli (1971).

[3] John Gooch, "Re-conquest and Suppression: Fascist Italy's Pacification of Libya and Ethiopia, 1922–1939," *Journal of Strategic Studies* 28, no. 6 (December 2005): 1006.

[4] Lucio Ceva, *Storia delle forze armate in Italia* (UTET, 1999), pp. 111–13.

results were often slim.[5] As so frequently in colonial warfare, Italian military superiority was overwhelming: 33,500 troops, modern equipment, and the logistical facilities for the pacification of a country with a population of just over one million.[6]

Despite a clear military and technical edge over the insurgent forces, the Italian political and military leadership faced a resistance strong enough to drag them into a long campaign. Brian Sullivan, in his study of the Italian military during the interwar period, considers that the learning curve of the Italians in knowing the enemy was slow: "What they lacked, despite 11 years of experience in Libya, was an understanding of their opponents."[7] It is worth noting that the Italians tried to narrow their cultural and sociological shortcomings and sometimes demonstrated finesse in their pacification efforts.[8] Sullivan proposes that the campaign could have been much shorter with more money invested, a proposal that is certainly debatable.[9] Finally, Marshal Pietro Badoglio, chief of the Supreme General Staff, and General Rodolfo Graziani, newly appointed vice governor of Cyrenaica, understood in the summer of 1930 that the guerrilla could be beaten only with an increased use of violence in the repression.[10] Their combined efforts in late 1930 and 1931 to isolate the insurgents from the population eloquently demonstrated the effectiveness of this form of counterinsurgency, especially with the use of internment camps.[11]

The pacification of Libya was successfully achieved in early 1932.[12] By then, the territory of Libya was under effective Italian control, and armed violence against the occupying force ceased. This achievement was a necessity "before Mussolini could strike out on his own."[13] Now Mussolini had a free hand to prepare his own campaign, one that would have been politically and militarily risky to undertake without this previous success in North Africa.

---

[5] Denis Mack Smith, *Mussolini* (Weidenfeld, 1993), p. 170.

[6] Brian R. Sullivan, "A Thirst for Glory: Mussolini, the Italian Military and the Fascist Regime, 1922–1936" (Ph.D. diss., Columbia University, 1984), p. 227.

[7] Ibid.

[8] Gooch, "Re-conquest and Suppression," pp. 1008, 1010.

[9] Sullivan, "A Thirst for Glory," p. 255.

[10] Giorgio Rochat and Giulio Massobrio, *Breve storia dell'esercito italiano dal 1861 al 1943* (Einaudi, 1978), p. 246.

[11] For a thorough and blunt explanation of the role of violence and brutalization in winning pacification campaigns, see Gil Merom, *How Democracies Lose Small Wars: State, Society, and the Failures of France in Algeria, Israel in Lebanon, and the United States in Vietnam* (New York: Cambridge University Press, 2003), pp. 33–47, especially the strategy of "isolation," pp. 38–41. On internment, see Nicola Labanca, "Italian Colonial Internment" in Ruth Ben-Ghiat and Mia Fuller, *Italian Colonialism* (Palgrave Macmillan, 2005), pp. 27–36.

[12] Romano Canosa, *Graziani. Il maresciallo d'Italia, dalla guerra d'Ethiopia alla Repubblica di Salò* (Mondadori, 2004), p. 75.

[13] MacGregor Knox, *Common Destiny: Dictatorship, Foreign Policy, and War in Fascist Italy and Nazi Germany* (New York: Cambridge University Press, 2000), p. 87.

Even if it is difficult to evaluate the impact of the Libyan campaign on the Ethiopian one, Sullivan did not hesitate, stating

> Ruthlessness against civilians and insurgents, either in Europe or Africa, was not a novelty for the Italian military. But the particular techniques developed in Libya: that is, the destruction of food supplies, the massive use of poison gas, the institution of concentration camps and the resort to genocide, would all be applied later, on a massive scale, in Ethiopia. In addition, the same leaders, trained in Libya, would direct these measures.[14]

The invasion of Ethiopia was "the last campaign of colonial conquest to be fought by a European power."[15] According to Giorgio Rochat, a leading Italian military historian, it was a "national war," a war fought as much for the needs of internal politics as for those of international prestige, with massive military deployment, sophisticated and well-orchestrated propaganda, and an economic effort far beyond the necessities of a normal colonial war.[16] The Ethiopian war, fought between October 1935 and May 1936, became Mussolini's greatest success and maybe was the most popular war in the history of Italy.[17] The regime had its finest hour, and few people in Italy were contesting the achievement of *Il Duce*, even if popular enthusiasm did not rise instantly in October 1935.[18] The response of the international community and the economic sanctions against their country persuaded the Italians that this was a legitimate war.[19] Most thought that the well-being of the nation, its right to belong to the "great powers," and hence its colonial policy were as legitimate as the ones of Great Britain and France. The sanctions were then seen as a form of aggression against the Italian nation.[20]

The Italian invasion of Ethiopia was a huge military enterprise.[21] In this "American" type of war, logistical nightmares became more serious obstacles than the enemy's resistance.[22] After the initial advance into Ethiopian territory

---

[14] Sullivan, "A Thirst for Glory," p. 256.

[15] Michael Howard, "The Military Factor in European Expansion," in Hedley Bull, Adam Watson, *The Expansion of International Society* (Clarendon Press, 1984), p. 41.

[16] Giorgio Rochat, *Les guerres italiennes en Libye et en Éthiopie, 1921-1939*, Service historique de l'Armée de l'air (Vincennes, 1994), pp. 14–15. The war of 1911–1912 against the Ottoman Empire was also a national war, according to Rochat and to Oreste Bovio, *Storia dell'esercito italiano, 1861–1990*, Stato maggiore dell'esercito, Ufficio storico (Rome, 1996), p. 189.

[17] Ceva, *Storia delle forze armate*, p. 232.

[18] Smith, *Mussolini*, p. 197.

[19] Ibid.; MacGregor Knox, *Hitler's Italian Allies. Royal Armed Forces, Fascist Regime, and the War of 1940–1943* (New York: Cambridge University Press, 2000), p. 11.

[20] Pierre Milza and Serge Berstein, *Le fascisme italien, 1919–1945* (Éditions du Seuil, 1980), pp. 341–42.

[21] Ceva, *Storia delle forze armate*, pp. 234–35; Bovio, *Storia dell'esercito italiano*, p. 281.

[22] Ceva, *Storia delle forze armate*, p. 235. On the logistics of the Ethiopian campaign, Ferrucio Botti, *La logistica dell'esercito italiano (1831–1981)*, vol. 3, *Stato maggiore dell'esercito, Ufficio storico* (Rome, 1994), pp. 552–647.

in October 1935, the shortcomings of the minister of the colonies, Emilio De Bono, as commander in chief and the Duce's pressure on him put the Italian forces in a state of crisis.[23] Badoglio replaced De Bono by mid-November, and after some months of reorganization (and some Ethiopian successes), took the offensive in February 1936. He finally reached and occupied the capital, Addis Ababa, by 5 May. The war, it was said or thought, was over.

But it was not. Gooch reminds us that the victory was an illusion.

> Only one third of Ethiopia had been occupied, the Italians controlling the routes from Eritrea to Addis Ababa and from Somalia via Harar to Dira Dawa. Italy spent the next five years attempting to conquer the rest of the country, in which some 25,000 rebels were under arms in any one year, but large areas of the north and north-west permanently eluded their rule. To do this, Graziani and its successor, the Duke of Aosta, had 466,000 white and colonial troops in 1936, 237,000 (of whom three-quarters were white) in 1937, and 280,000 (of whom three-sevenths were white) in 1938–1939.[24]

It is difficult to know to what extent Mussolini understood what would follow the capture of Addis Ababa. "On the surface all appeared calm, but despite Mussolini's optimistic forecasts, the war was anything but over."[25] In the years to come, the pacification process cost Italy 9,555 dead and 140,000 wounded or sick.[26] Italian forces in East Africa were in a position of strategic weakness; it was easy for the British to "raise the tribes" against the Italian troops in 1940–1941.[27] The differences between the Libyan success and the Ethiopian failure in pacification are substantial:

> The results are explicable in large part as the consequence of differences in the nature and effectiveness of operational methods; in Libya, the Italians had the time to experiment and the military capacity to develop the expertise required to win their war, whereas in Abyssinia neither was the case. Physical geography was an important factor in deciding the outcome in both theatres. So, too, was political ethnography and the uses which the Italians did or did not make of it. Finally, the harshly repressive policies adopted in Libya and in Abyssinia were pursued with self-consciously Fascist rigour, with quite different consequences.[28]

Despite hundreds of thousands of troops, equipment, air power, use of gas, and brutal and ferocious attempts to destroy the insurgency, Fascist Italy

---

[23] Ceva, *Storia delle forze armate*, p. 235.

[24] Gooch, "Re-conquest and Suppression," p. 1022.

[25] Angelo Del Boca, *The Ethiopian War, 1935–1941*, trans. P. D. Cummins (Chicago: University of Chicago Press, 1969), p. 212.

[26] Gooch, "Re-conquest and Suppression," p. 1025.

[27] Dawn M. Miller, "Raising the Tribes: British Policy in Italian East Africa, 1938–41," *Journal of Strategic Studies* 22, no. 1 (March 1999): 96–123.

[28] Gooch, "Re-conquest and Suppression," p. 1006.

was not able to overcome the Ethiopian resistance.[29] It seems that Ethiopia became a place where violence and brutality didn't work.[30]

Mussolini's Blindness

To pacify a country is a complex and difficult task. Political shrewdness is a prerequisite, not an option. Whatever the origins of Mussolini's dream of rebuilding the Roman Empire, he clearly misunderstood the implications of conquering Ethiopia. His blindness—his incapacity to see what should be done or what the reality truly was—operated at two levels.

A first level of blindness was grounded in his belief, conscious or not, that victory came with the fall of Addis Ababa. Two reasons probably explain why Mussolini associated the fall of the capital with victory. The first was linked to the necessities of propaganda, internal and external, in a time of European political tension. Mussolini needed a rapid decision, and the capture of Addis Ababa became the decisive tool of his propaganda apparatus: Italy won a campaign over an enemy in a war that many European observers had predicted would be a long war.[31] With this victory, the debate about the legitimacy of the invasion became irrelevant. Italy had to be recognized as a world power, its Duce as a great leader. The second reason had a more technical nature: the capture of the capital city was also marking the relative disintegration of the Ethiopian army. Numerous Ethiopian soldiers were still armed but leaderless, and most units lost their cohesion. If we also consider that the capture of a capital was frequently seen in the history of conventional warfare as a decisive step, if not *the* decisive step, of a military campaign, the temptation was too strong for Mussolini to resist. The fighting was over, and it was thought that the Ethiopians had accepted Italian rule. The idea that armed resistance against the invader would now take a different form did not seem to have occurred to Mussolini.

These two reasons are not sufficient to excuse Mussolini for his blindness. His incapacity to understand the nature of war was deeply rooted in his personality.[32] If war is, to paraphrase Quincy Wright, "a violent conflict between two organized human groups," then the fall of Addis Ababa was not the end of the war, but surely the opening of Pandora's box.[33] For Mussolini,

---

[29] On the use of gas during the war and the pacification campaign, Alberto Sbacchi, "Poison Gas and Atrocities in the Italo-Ethiopian War (1935–1936)," in Ben-Ghiat and Fuller, *Italian Colonialism*, pp. 47–56, and Rochat, *Les guerres italiennes en Libye et en Éthiopie,* pp. 183–226.

[30] For many examples of the opposite result, Merom, *How Democracies Lose Small Wars*, pp. 33–47.

[31] Smith, *Mussolini*, p. 201.

[32] On this topic, see Renzo De Felice, *Mussolini l'alleato. I. L'Italia in guerra 1940–1943, 1. Dalla guerra "breve" alla guerra lunga* (Einaudi, 1990), p. 52.

[33] Quincy Wright, *A Study of War*, 2d ed., abridged (Chicago: University of Chicago Press, 1983), pp. 6–7.

the objective of the war was achieved with the fall of the city; he didn't foresee what was to come, probably because he didn't care. Call it irregular war, nonconventional war, guerrilla war, or small war, the reality was that from mid-May 1936 until the beginning of the Second World War in East Africa in 1940, Italian troops and Ethiopian fighters were in a state of war.[34] The prestige and the glory gained by the victory vanished quite rapidly. In the days after the capture of the capital,

> The standing force of 426 officers and 9,934 men that garrisoned the forest city and its extensive outskirts was, in fact, almost ringed round by 50,000 leaderless Ethiopian soldiers, the majority of whom had been disbanded, but all of whom were armed. Most of the empire, indeed, was only nominally under Italian domination.[35]

Libya should have been a lesson learned: a national war turning into a long pacification campaign. But Ethiopia was of much greater complexity than Libya: a large population, a difficult geography, and, most of all, a civilization that would not accept foreign domination without fighting. Mussolini should have known that. If there was some preparation and planning for the invasion, there was none for its pacification.[36] Mussolini's blindness and military amateurship are the main reasons for that. Considering his *antimilitarismo*, his military entourage lacked the necessary influence and power to make him understand the implications of pacification.

His blindness operated at a second level. Time is among the most precious weapons in a pacification campaign. It is undeniable that "oppressors hardly ever intended to let insurgency wars drag on or bleed them so much as to make their losses unacceptable."[37] But the more time a conqueror has, the more he can figure out what is the right strategy to win against an insurgent movement.[38] Here lies a major difference between Libya and Ethiopia. In the first case, Mussolini was in no hurry; he inherited a situation at a time of relative European political tranquility. The pacification campaign in Libya lasted almost ten years and was characterized by major differences in the areas to be controlled, Tripolitania and Cyrenaica, and by a necessity to adapt to these differences.[39] Despite the length of the process, the Italian military was never truly pressured by political necessities or international circumstances. Time was on its side

[34] On terminology, see Frank G. Hoffman, "Small Wars Revisited: The United States and Nontraditional Wars," *Journal of Strategic Studies* 28, no. 6 (December 2005): 915–16.

[35] Del Boca, *The Ethiopian War*, p. 212.

[36] Rochat, *Les guerres italiennes en Libye et en Éthiopie*, p. 230.

[37] Merom, *How Democracies Lose Small Wars*, p. 34.

[38] On these strategies, ibid., pp. 33–47.

[39] On the many difficult steps of adaptation, Gooch, "Re-conquest and Suppression," pp. 1007–21; Rochat and Massobrio, *Breve storia dell'esercito italiano,* p. 247.

and adaptation became possible. The *esercito italiano* finally summoned the flexibility and determination it needed to perform efficiently.[40]

The invasion of Ethiopia was made possible only "by the rupture of the European equilibrium caused by German rearmament."[41] Mussolini needed a quick victory because he became prisoner of this quite volatile European equilibrium. On 24 October 1936, almost seven months after the glorious days of May and the fall of the capital, he declared: "It took us seven months to conquer the empire, but to occupy and pacify it will take us far less time."[42] It is impossible to know if Mussolini really believed what he said and if he understood the complexity of the situation. But most probably, he had no interest in the management of the pacification of the newly acquired empire. European politics diverted his interest from Africa. The Spanish Civil War became his next political and military adventure, and his objective, again, was to establish his personal prestige as a great leader.

The pacification of Ethiopia would have needed a well-prepared strategy and, probably, more resources for the *forze armate* than the ones allocated after 1936. None of these conditions were met. European politics recaptured the attention of Mussolini, and improvisation in repression became the only strategy used by Graziani. But most of all, time was short and the timing was bad. The probability of a European war grew rapidly, and the unpacified East African Italian empire became more of a burden than an asset for a future war, a situation perfectly acknowledged by the Duke of Aosta in 1939. Mussolini didn't give time a chance and then considerably reduced the possibilities of adaptation to the insurgency.

Mussolini was responsible for not seeing the rise of an insurgency. He was also guilty of not giving his "grand colonial policy" the attention, the resources, and, most of all, the amount of time required to make pacification possible. His thirst for glory and desire to affirm himself as a great ruler largely diminished the chances of conquering the Ethiopian territory. But there were also contingencies. At least three made the pacification of Ethiopia a desperate undertaking.

Contingency One

As soon as the military campaign in Ethiopia was over, the Duce dictated a policy of direct rule over the country. In a series of famous telegrams, Mussolini declared that Fascist Italy would not, in any way, share power with the local chiefs or nobles. "No power to the *ras*" became Mussolini's strategy for the control and management of Ethiopia. The sources of this policy are unknown. Gooch noted that:

---

[40] Rochat and Massobrio, *Breve storia dell'esercito italiano,* p. 247.

[41] Ceva, *Storia delle forze armate*, p. 232.

[42] Del Boca, *The Ethiopian War,* p. 216.

> Mussolini, however, refused to contemplate it for reasons that he never made explicit but which amounted to a conviction that for historical, ideological and practical reasons Italian power in Ethiopia was not to be shared with natives and not to be seen as shared.[43]

This policy was a contingency, and in no way something that had to happen. Mussolini's political unpredictability could have led to a different policy. Rochat proposed that the Libyan experience was the reason Mussolini was obsessed by the idea of depriving the ruling class of any form of political power.[44] Mussolini thought that the experience of collaborating with Libyan elites made the pacification process longer. Even if Rochat's proposal seems logical, it is impossible to demonstrate if, indeed, this was the case. What is known for sure is that the use of extreme violence and brutalization in a country with a very small population destroyed the Libyan insurgency in 1931–1932.

If the policy of no power to the *ras* was an occurrence preordained, then Badoglio and Graziani should have seen the coming of its implementation. Needless to say, neither of them was consulted; Badoglio disagreed openly and Graziani expressed numerous reservations, in vain.[45] The strong support of Alessandro Lessona, the new minister of colonies, transformed the policy into a slogan, a catchword used at every occasion to justify indiscriminate violence. According to Rochat, Lessona supported the policy of no power to the *ras* in order to consolidate the power of his ministry over the Empire: any sharing of power would have meant less for him.[46] Interestingly enough, Graziani did talk frequently with nobles and chiefs who were opposed to the previous political order and tried to use them as much as possible as supporting elements of the Italian rule. At the end of July 1936, his proposal to give to some chiefs and nobles some sort of power met Mussolini's refusal.[47] The Duce's obsession with a policy of direct rule was not only against common sense, it also became one of the reasons why the hearts and minds of the Ethiopian people were never conquered.

Contingency Two

The assassination attempt on Graziani's life on 19 February 1937 was certainly a major element in unleashing the violence and brutalization that were the trademarks of the Italian occupation of Ethiopia. In the days and months following the attempt, violence and brutalization took a different dimension; they became almost systematic and much more extended. Rochat carefully talks

---

[43] Gooch, "Re-conquest and Suppression," p. 1026.

[44] Rochat, *Les guerres italiennes en Libye et en Éthiopie,* p. 231.

[45] Ibid.

[46] Ibid., p. 233.

[47] Ibid.

of a "genocide" process.[48] Despite the fact that violence and brutalization have been used extensively in counterinsurgency strategies in history, it has been proposed earlier that violence and brutalization became ineffective tools in the pacification of the freshly conquered empire. The Italian failure to pacifying Ethiopia was caused, in part, by the counterproductive effects of violence and brutalization, both largely deriving from this second contingency.

The failed attempt was a contingency, something that could have never taken place. The two individuals involved were not Ethiopians. No large groups were really supporting them, despite many attempts by Fascist authorities to link specific groups to the action. In fact, it was almost an isolated action. But the "orgy of violence" that followed was real and degenerative. At the news of the attempt, Guido Cortese, a Fascist party leader in the capital, ordered a vendetta. The perpetrators were ordinary Italian colonists; the victims, Ethiopians who were living in the poorest neighborhoods of the city. This wave of killing was not planned; it was an instantaneous reaction, but one that received the support of Graziani, Lessona, and, of course, Mussolini. After the terrible days of killing in the capital, a more extended operation took place in order to identify and kill all the people who could have been linked to the assassination attempt. Eventually, all the leading social and political elite of Ethiopia became the target of a repression campaign, the killing of monks at the monastery of Debra Libanos in May being only one example. In this campaign of terror, Graziani asked for the deportation of many notables, the destruction of entire parts of the capital, and the installation of a massive concentration camp for those rendered homeless. Rochat says that the last request is evidence that Graziani lost his nerve, and, this time, even Mussolini showed some restraint.[49]

In the months following the attempt, several thousand Ethiopians of different social statuses were killed in reprisals, and the failed attempt against the viceroy opened a gap between the Italian authorities and the local population.[50]

> If the oppressors are uninterested in reconciling their interests with those of the oppressed, then the incentive to escalate the level of violence is compelling. The chances are that a less selective use of violence will cut the costs and reduce the time of planning and executing each of the strategies of pacification. From an expedient point of view, then, the movement on the strategic scale from selective eradication to indiscriminate annihilation is tempting. In that sense, counterinsurgency is inherently degenerative.[51]

This gap did not narrow as violence and brutalization became more than ever the sole tool of pacification. Graziani lost his position in November 1937.

---

[48] Ibid., p. 263.
[49] Ibid., p. 258.
[50] Canosa, *Graziani. Il mareciallo d'Italia,* p. 159.
[51] Merom, *How Democracies Lose Small Wars*, pp. 46–47.

Ethiopia proved to be a more complex challenge for him than Libya had been. The chances that his replacement would make a difference were slim.

Contingency Three

Amedeo di Savoia, Duke of Aosta, became the new viceroy of Ethiopia in late 1937. The man was seen inside the military leadership as a moderate and one who could change the tide even if, because of Mussolini's blindness, time was running out.[52] As Del Boca noted, expectations were high.

> It had been impressed on him that he was to pacify the country at all costs but he was to achieve this end by more humane, more conciliatory methods than his predecessor. The Duke, a cultured and liberal-minded man, was by nature ideally equipped for the task of persuading the Ethiopians to cooperate with the Italians, but the errors made by Graziani were irreparable.[53]

If Mussolini was so persuaded that only repression, violence, and the policy of direct rule would make pacification and control of the empire possible, then the choice of the Duke of Aosta is a contradictory one. But this paper proposes that slogans like "all rebels made prisoner are to be shot" and "no power to the *ras*" were far from being the result of a serious decision-making process based on the analysis of the Ethiopian situation. The choice of di Savoia was a contingency, one possibility among others. The probability that he was deliberately chosen for his liberal-minded attitude and his moderation is low.

The new viceroy gave a series of orders dedicated to imposing some restraint upon the use of violence by Italian forces.[54] Soon after his arrival, he got a new commander in chief of the armed forces in Italian East Africa, General Hugo Cavallero. Cavallero's knowledge of Ethiopia was superficial in every sense, his expertise in colonial warfare or insurgency thin. His presence, a complementary contingency, complicated the task of the viceroy. In 1938, Cavallero launched a series of military operations against the "patriots," claiming the deaths of thousands of them in many of the most rebellious regions of the country.[55] Cavallero continued the brutal methods used by Graziani, particularly the use of poison gas.[56]

The opposing personalities and numerous disagreements between the Duke of Aosta and Cavallero made the task of pacification no easier and slowed the former's attempts to try different methods. Cavallero resigned in May 1939, and it was only then that the duke was able to implement a new policy

---

[52] Gooch, "Re-conquest and Suppression," p. 1025.
[53] Del Boca, *The Ethiopian War,* p. 239.
[54] Ibid., pp. 241, 248.
[55] Gooch, "Re-conquest and Suppression," p. 1025.
[56] Ibid.

that was a partial rupture with Mussolini's no power to the *ras*.[57] The viceroy recognized that some remote areas were de facto outside the Italian occupation, and tacitly left them to the insurgents' control.[58] In other areas, where cooperation between the Italians and the locals was encouraged, the results obtained were good but not sufficient to completely defeat the insurgents. They came too late, and the coming of the Second World War gave a second life to the insurgency in 1940.

Conclusion

It is tempting to propose that the Italian adventure in Ethiopia was a fiasco. The military victory announced with the fall of Addis Ababa and Haile Selassie's exile was a delusive one. The pacification campaign cannot be considered a success. The empire-building strategy was a failure.[59] Almost five years after the emperor's exile to London in 1936, he was back in his liberated country. Mussolini's Roman Empire was gone.

Adaptation over time, flexibility, and extreme brutalization made the pacification of Libya a success for Fascist Italy. Merom's theory is that

> Violence is not only the primary means of getting the desired results of war. Rather, it is also a way of managing its costs. In other words, states resort to greater and less selective methods of brutality in pacification wars not only because these prove to be effective, but also because they prove to be efficient. Higher levels of violence can cut down on the investment and loss of manpower and material, both through the destruction involved and the fear generated.[60]

This theory was proved to be inaccurate in the case of Ethiopia. Indeed, extensive use of violence and brutalization probably aroused more defiance than defeatism. Mussolini's misunderstanding of the consequences of his invasion of Ethiopia put Italy into a complex campaign of pacification. His obsession to play a role in European politics, particularly in Spain, diverted him from African affairs. The pacification of Ethiopia needed time, certainly as much as in the Libyan case. But because of Mussolini's miscalculations, time rapidly ran out as the pacification efforts bogged down into a stalemate in 1938–1939. Mussolini's role in the failure to pacify Ethiopia is not a debatable issue, although it is not the entire explanation. The three contingencies developed in this paper show that sometimes historical explanations are to be found through occurrences free from necessity or preordination. The "no power to

---

[57] Rochat, *Les guerres italiennes en Libye et en Éthiopie*, p. 34.

[58] Ibid.

[59] Haile Larebo, "Empire Building and Its Limitations: Ethiopia (1935–1941)," in Ben-Ghiat and Fuller, *Italian Colonialism*, pp. 83–94.

[60] Merom, *How Democracies Lose Small Wars*, p. 43.

the *ras*" approach, the failed attempt on Graziani's life, and the unfortunate pairing of the Duke of Aosta and Cavallero made the pacification of Ethiopia almost an impossible task.

The Roots of Dutch Counterinsurgency Balancing and Integrating Military and Civilian Efforts from Aceh to Uruzgan

Thijs W. Brocades Zaalberg

At the outset of the twenty-first century, the United States is obviously not the only Western country troubled by the harsh realities of counterinsurgency campaigning. Several of its closest allies, such as the United Kingdom, Canada, Australia, and the Netherlands, have become embroiled in this form of irregular warfare. On two occasions, in both Iraq and Afghanistan, the Netherlands armed forces followed in the wake of a U.S.-led offensive in order to help stabilize these nations confronted with chaos and insurgency. From July 2003 until March 2005, some 1,200 Dutch troops were in charge of stabilizing Al Muthanna Province in southern Iraq. The Dutch operation in the south, executed under British divisional command, had some of the characteristics of a counterinsurgency campaign, but a full-blown insurgency did not emerge at the time in Al Muthanna. Therefore, this Dutch operation cannot be considered a counterinsurgency operation as such.[1] However, in Afghanistan's southern Uruzgan Province, over 1,500 Dutch troops have been engaged in a complex counterinsurgency operation since the summer of 2006 as part of NATO's International Security and Assistance Force (ISAF).[2]

In the colonial days, the Dutch would have referred to these operations as pacification campaigns. Unhampered by elaborate modern definitions of counterinsurgency, irregular warfare, or stabilization and reconstruction, they may have also used the generic term *small wars*. This term, a direct translation of the term *guerrilla*, was in vogue in the late nineteenth and early twentieth century, during the era of modern imperialism. At the time, the Netherlands was a minor power, but it owned vast colonial possessions. The Netherlands East Indies, the precursor to the state known after 1949 as Indonesia, was created through gradual imperial conquest since the seventeenth century. The Dutch began with wresting a small foothold on Java in 1619, at a time when the Dutch Republic still ranked among the world's greatest powers.

[1] In January 2006, Al Muthanna was the first Iraqi province outside Kurdish territory where the Iraqi government gained full responsibility for internal security. The Netherlands Institute for Military History (NIMH) has started research for a publication on Dutch military operations in southern Iraq.

[2] In December 2007, the Dutch government committed itself to keeping this force in Uruzgan until mid-2010.

They finalized the borders of the state of Indonesia as it currently is with the conquest of Aceh (northernmost Sumatra) and Bali in 1909. On the eve of the Second World War, only British India competed with the Netherlands East Indies in the wealth it brought to a colonial power. The archipelago, with a population estimated at seventy million, was certainly far more important to the Dutch economy than India was to Britain's.

During this process of colonial expansion, the Dutch colonial army frequently met with irregular resistance. Right after the Second World War, the Dutch armed forces again found themselves fighting a guerrilla war during the Indonesian struggle for independence that lasted until 1949. Little is known internationally about the Dutch experience in small wars and counterinsurgency. Professor Ian F. W. Beckett appears to be the only non-Dutch author who, in an English publication, briefly mentions the Dutch experience in his comparative analysis called *The Roots of Counter-Insurgency*.[3] But there is another important reason for analyzing Dutch experience in countering insurgencies. The Dutch style of operations in Afghanistan's Uruzgan Province has placed a lot of emphasis on avoiding the use of military force—or what in modern military jargon is awkwardly called nonkinetic methods. According to Beckett, "A particular army's counter-insurgency practice has so frequently evolved from its past colonial experience."[4] The question this paper therefore addresses is whether there is continuity in the Dutch approach to fighting irregular opponents. Is there a tradition—perhaps similar to that of the British with their minimum-force philosophy—that can explain the current Dutch approach to countering irregular opponents?

A "Dutch Approach" in Afghanistan?

NATO operations in southern Afghanistan neatly match current definitions of *counterinsurgency*. However, the Dutch government officially avoids the term in relation to its contribution to this mission. An American audience is likely to ask what these military forces are doing in southern Afghanistan if not countering the Taliban insurgency. It appears to be a matter of emphasis. At the outset of the mission, the Netherlands government tended to present

[3] Ian Beckett, *The Roots of Counter-Insurgency: Armies and Guerrilla Warfare, 1900–1945* (New York: Blandford Press, 1988), p. 153. Another comparative analysis, one that fails to mention the Dutch experience, is David Charters and Maurice Tugwell, eds., *Armies in Low-Intensity Conflict: A Comparative Analysis* (London: Brassey's, 1989). There are many publications in English on the Indonesian revolution and the Dutch political and military response but not from a counterinsurgency perspective. Petra Groen has written an excellent dissertation on Dutch military strategy in Indonesia between 1945 and 1949, and many other Dutch studies on military operations exist, but, even in the Netherlands, little is known about policing and the civil administrative side of the struggle. Petra Groen, *Marsroutes en Dwaalsporen: Het Nederlandse Militair-Strategische Beleid in Indonesië* (Den Haag, 1991).

[4] Ian F. W. Beckett, "Forward to the Past: Reflections on British Responses to Insurgency," *Militaire Spectator* 177, no. 3 (March 2008).

Dutch operations in Uruzgan as a reconstruction effort rather than irregular warfare. Stabilization operations by military forces were only a part of the solution—"enablers" for the real mission: reconstruction of a war-torn society. After exchanging ideas with senior European officers in southern Afghanistan in late 2006, the Australian counterinsurgency specialist and adviser David Kilcullen, slightly uncomfortable with what he heard, called this "a development model to counter-insurgency."[5]

In line with the emphasis on stabilization and reconstruction rather than combat, the Dutch Ministries of Defence and Foreign Affairs stressed the importance of "the comprehensive approach," the integrated civil-military approach, or "3-D approach," tying together defense, diplomacy, and development. In early 2006, there was even talk in political and military circles in The Hague about a certain unique "Dutch approach" to stabilization and reconstruction that placed extra emphasis on respect for the local population and its customs.[6] From a historical perspective, the latter term seemed rather pretentious. If the direct approach to counterinsurgency is a singular focus on the annihilation of enemy forces with military means, the Dutch—both politicians and the military—embraced an indirect, population-centered approach to countering insurgents, which is by no means unique.

One could raise doubts about political rhetoric matching operational realities on the ground in Afghanistan. The 1,500-strong Dutch military force in southern Afghanistan included a battle group, Special Forces, and tracked 155-mm. howitzers, with Dutch Apache attack helicopters and F–16s in support. The Provincial Reconstruction Team (PRT) was officially at the heart of the mission but did not exceed fifty persons.[7] However, even in military circles, little emphasis was placed on the use of military force. The population, or rather, "the hearts and minds of the people," was considered the center of gravity, rather than killing or capturing the insurgents. In April 2007, the *New York Times* quoted the commander of the Dutch Task Force Uruzgan, Col. Hans van Griensven, in an article entitled "Dutch Forces Stress Restraint in Afghanistan." The colonel stated, "We are not here to fight the Taliban. We are here to make the Taliban irrelevant."[8] When it came to applying some of the classic British counterinsurgency principles, the Dutch appeared to be holier

---

[5] Kilcullen quote in George Packer, "Knowing the Enemy: Can Social Scientists Redefine the 'War on Terror'?" *New Yorker*, 18 Dec 2006.

[6] On 9 January 2006, the term *Dutch approach* was first officially coined in relation to the mission in Uruzgan by Chief of Defence General Dick Berlijn. Steven Derix, "Zonder troepen blijft de nodige hulp uit: hoogste militair pleit voor missie Uruzgan," *NRC Handelsblad*, 10 Jan 2006. See also Robert H. E. Gooren, "Soldiering in Unfamiliar Places: The Dutch Approach," *Military Review* (March–April 2006).

[7] Task Force Uruzgan (TFU) was officially "built around" the PRT. Brief aan de Eerste en Tweede Kamer van minister van Buitenlandse Zaken en de minister van Defensie betreffende Nederlandse bijdrage aan ISAF in Zuid-Afghanistan, 22 Dec 2005 (DVB/CV-388/05).

[8] C. J. Chivers, "Dutch Forces Stress Restraint in Afghanistan," *New York Times*, 5 Apr 2007.

than the Pope. These "classic counterinsurgency principles" include stressing the need for a political rather than a military solution, civil-military cooperation, "winning the hearts and minds," minimum use of force, and so on.[9]

To answer the question of whether there is continuity to the Dutch history of counterinsurgency operations, this paper harkens back to the two largest and most crucial campaigns in our colonial past: the war in Aceh (1873–1909) and the Indonesian War of Decolonization (1945–1949). In order to allow me to make such broad statements on more than 130 years of history, this paper focuses on one element of counterinsurgency campaigning: the difficulty involved in balancing and tying together civil and military efforts. This element is at the heart of successful modern counterinsurgency operations and appears to be the Achilles heel of current Western interventions. A closely related issue is, of course, the use of minimum or measured military force. This paper focuses on military practice rather than doctrine, since doctrine and handbooks played a minor role during the actual Dutch campaigns, then as well as now. However, the ability to become what John Nagl calls a "learning institution" brought the Dutch colonial army its self-declared moments of glory. It managed to succeed where the Indonesian government and army never succeeded: defeating an insurgency in Aceh. But it took them more then thirty years.

Aceh, 1873–1909

Most Western governments and armies have a terrible record when it comes to learning from counterinsurgency experience. The nineteenth-century Dutch colonial army in the East Indies is a case in point. At the outset of the Java War in the 1820s and every new campaign since, the army advanced in large columns of heavily armed forces, including cavalry and artillery. In campaigns that lasted many years, the army often searched for an elusive enemy and a decisive battle to win. Those indigenous forces that chose to fight conventionally were mostly defeated with ease, but irregular opponents continuously hampered colonial ambitions. Lessons were, in the end, learned by some visionary commanders—those who adapted their organization and tactics to the enemy—but hard-won knowledge was quickly lost between campaigns.

The war that started in Aceh in 1873 was the largest, longest, and most vicious of all campaigns. For more than twenty years, the war against the Muslim Acehnese progressed disastrously for the Dutch. The conflict reached a deadlock by the 1880s, after two failed offensives along the lines just mentioned—the search for a decisive military victory. The colonial army held only a narrow defensive perimeter around the capital of the sultanate of Aceh. This "Dutch Dien Bien Phu," as one historian called it, came under frequent

[9] Beckett, *The Roots of Counter-Insurgency*, p. 12.

attack from fanatic irregulars.[10] The Dutch answer was punitive force that often directly targeted the population, which only stiffened the resistance to Dutch rule.

Only between 1898 and 1903 was the tide turned, partly by employing light and flexible forces, called the *Korps Marechaussee*. These constabulary-type units consisted mostly of Ambonese and Javanese troops led by Dutch officers. They also have been compared to Special Forces. An ambitious new commander, Col. J. B. Van Heutsz, initiated a steady and intensive pacification campaign aimed at controlling territory and the population. Thereto he employed intensive offensive patrolling by small "flying columns" and created a system of blockhouses. Van Heutsz, who would soon become a general, was advised by an authority on Islam, Dr. Christiaan Snouck Hurgronje. This prominent scholar had done extensive research in Acehnese society and became convinced that the enemy was waging an Islamic holy war, with radical clergymen at the heart of the resistance. Apart from being a cultural adviser, Snouck provided crucial political advice and intelligence.

The joint vision of these two men, the general and the scholar, was to break with the massive, "scuttle and burn–type" punitive force. Instead, they propagated a new form of "surgical force." To use their own terminology: give the Acehnese "a sensitive beating" and place "the foot on the neck." Coercion was needed, but the population was to be treated in a humane fashion. Surgical force was to be complemented by a prosperity policy aimed at winning over these "future subjects." By 1903, the backbone of Acehnese resistance was broken. Van Heutsz, by then a national hero, became the governor general of the entire Dutch East Indies. Decades of relative peace followed after the remnants of resistance were quashed in Aceh and elsewhere in the Indies by 1910. This situation lasted until the Japanese invasion in 1942. Aceh can be considered the birthplace of a new Dutch counterinsurgency method that would be used to great effect elsewhere in the archipelago.

On the basis of this information, one could conclude that the outlines of a more subtle, integrated "Dutch approach" toward counterinsurgency started to emerge. Last year, a Dutch battalion commander in Afghanistan was quoted in the press referring to Van Heutsz's methods as a source of inspiration.[11] Light and flexible units, measured force, an emphasis on cultural awareness,

[10] H. W. van den Doel, "Military Rule in the Netherlands East Indies," in *The Late Colonial State in Indonesia: Political and Economic Foundations of the Netherlands Indies 1880–1942*, ed. Robert Cribb (Leiden: KITLV Press, 1994), p. 62.

[11] "Als vlooien op een wilde hond: Oude Van Heutsz-strategie is inspiratie voor aanpak Taliban," *Elsevier*, 30 Dec 2006. The battalion commander whose unit carries the name and tradition of the Regiment Van Heutsz is likely to have been misquoted. Instead of referring to Van Heutsz's strategy as a whole as a "source of inspiration," he told the *Elsevier* reporter that he had looked, among others, at the Marechaussee tactics before heading for Uruzgan in the summer of 2006. He and his intelligence officer also studied the U.S. Marine Corps Combined Action Platoon (CAP) program in Vietnam and hoped to apply similar methods in Uruzgan. A correct quotation of the commander can be found in Noel van Bemmel, "Lessen uit Atjeh

and intelligence—it all seemed to be in line with what the Dutch nowadays want to be in Afghanistan. So can we conclude that there is continuity in the Dutch approach to fighting insurgencies? The answer is clearly negative. Such a conclusion would be based on a distorted picture of the reality on the ground in Aceh. As late as 1904, after the insurgency in the heartland of Aceh was essentially broken, a six-month campaign was undertaken by a large column into the hinterland of northern Sumatra. Acehnese villagers—men, women, and children—were killed on a massive scale during this particular operation. Between one-quarter and one-third of the population in this region perished.[12]

Although unmatched in brutality, this massacre by Maj. Gotfried Coenraad E. van Daalen was not just an incident. During the crucial period 1898–1903, as well as the following six years, the campaign under Van Heutsz as a whole relied on brute military force rather than enlightened methods. Instead of winning over the population, military administrators exercised draconian control over the entire population. They inflicted severe collective punishment and fines on entire communities suspected of supporting the insurgents. No one other than the infamous Van Daalen succeeded Van Heutsz as the military governor of Aceh. In his hands, the policy of prosperity for the Acehnese people came to nothing.

Obviously, theory and practice did not match. The "Dutch approach" to colonial expansion and pacification in Aceh in this period relied largely on brute force and was highly militarized. In 1904, just after the news of the brutalities had broken in the Netherlands, a Dutch member of parliament complained, "What worries me at this point is that we are moving towards a militaristic atmosphere. . . . We have a soldier for Colonial Minister, a soldier for Governor-General [of the Netherlands East Indies], a soldier for Governor of Aceh."[13] He could have continued. Also on the district and municipal level, military administrators were often in charge in the more volatile outer areas. Dutch civil administrators were scarce and used only sporadically, and then only in an advisory role. A shortage of willing and able civilian administrators was one reason for the militarized approach, but there was one other. The difficult pacification campaign required extremely close coordination of government and military operations, and the easiest way to achieve this was to give military officers comprehensive civil powers. In the Netherlands East Indies, civil and military powers were united under military command far more often than in British India or with the French in their colonies. This militarized colonial administrative concept was designed for crises such as in Aceh after

voor Uruzgan: Hernieuwde Interesse voor Innovative maar ook Keiharde Campagne," *De Volkskrant*, 12 Nov 2007.

[12] Paul van 't Veer, *De Atjeh-oorlog* (Amsterdam, 1969), p. 269.

[13] Martin Bossenbroek, *Holland of Zijn Breedst: Indië en Zuid-Afrika in de Nederlandse Cultuur Omstreeks 1900* (Amsterdam, 1996), pp. 44–45.

1880. However, it would be maintained in many areas outside Java until the Japanese occupation in 1942.[14]

War of Decolonization, 1945–1949

So what did the Dutch learn from the Aceh experience in the late 1940s, when they tried to reoccupy their colony after the Second World War? They learned one lesson. The reoccupation of the Indonesian archipelago was the number-one national priority after 1945, but the Netherlands did not try to reconquer Aceh. Overall, however, old habits died hard. The Netherlands tried to combine political, diplomatic, and military measures, but ultimately military force and repression became the key instrument of Dutch policy.[15] This may have worked during the war in Aceh, an internationally isolated and geographically limited conflict that took place around the turn of the century. However, it did not work against a nationwide nationalist movement that drew broad support from the population and attracted attention from all over the world. The Indonesian people, seventy million strong at the time, had just seen the Dutch being defeated with ease by the Japanese Army. The Dutch had tumbled from their imperial pedestal and, against the odds, tried to climb back on.

What was the method used? After 1945, the Dutch sought a conventional military solution to the Indonesian nationalist revolt by twice relying on a speedy military offensive. These offensives, euphemistically called police actions, were highly successful in conventional military terms. During the offensive in July and August 1947, the Dutch secured the key economic objectives. During the second action in December 1948 and January 1949, the Dutch armed forces successfully captured the republican "rebel" capital Yogyakarta and even succeeded in arresting key nationalist leaders, including the Republic's President Sukarno. However, both offensives caused widespread international indignation. They also left the Dutch with immense territories and a massive population to control. Without a proper counterinsurgency strategy, the Dutch failed during the costly and, at times, brutal pacification campaign in the countryside.[16]

Measuring the Dutch performance against the six classic counterinsurgency principles provides a further explanation for the failure of the militarily superior

[14] H. W. van den Doel, "De ontwikkeling van het militair bestuur in Nederlands-Indie: de officier-civiel gezaghebber, 1880–1942," *Mededelingen van de Sectie Militaire Geschiedenis*, nr. 12 (Den Haag, 1989), pp. 29, 48. For an English version of this article, see Van den Doel, "Military Rule in the Netherlands Indies."

[15] Petra Groen, "Militant Response: The Dutch Use of Military Force and the Decolonization of the Dutch East Indies, 1945–1950," *Journal of Imperial and Commonwealth History* 21, no. 3 (September 1993): 30.

[16] Petra Groen, *Marsroutes en Dwaalsporen: Het Nederlandse Militair-Strategische Beleid in Indonesië* (Den Haag, 1991).

Dutch to defeat the insurgency. These principles emerged from the British colonial experience and culminated in the British answer to the Malayan Emergency. Many of these same principles can be found in the U.S. Marine Corps *Small Wars Manual*. The Dutch failed on all six accounts. First, they failed to ensure political primacy, a realistic political aim, and relied on military force instead. Like the French in Vietnam and Algeria, they ignored that decolonization after the Second World War was inevitable. Second, they failed to stick to the principle of measured use of force. Artillery and airpower were often used indiscriminately. Although not on a scale comparable to the French in Algeria, summary justice and third-degree interrogation was applied, particularly by Special Forces and military intelligence personnel. Third, they failed to create a successful and balanced mechanism for civil-military cooperation among the military, civil administration, and police. Military rule increasingly became the norm. Fourth, the Dutch failed to separate the insurgents from the population. There was neither a successful "hearts and minds" campaign, nor did the colonial government succeed in effectively controlling the population. Fifth, Dutch intelligence on the enemy was poor. Despite many years of colonial rule and linguistic and cultural knowledge, we did not know what drove our enemy, what the strength of the nationalist movement was, and what the population wanted. Dutch military leaders convinced political leaders that the decapitation of the insurgency, the arrest of its leaders, and the occupation of the rebel capital Yogyakarta would do the job. Finally, the Dutch lacked patience. Hoping for quick results, we focused on two speedy offensives instead of progressive pacification and long-term reform.

Let us return to the third point: civil-military balance and integration. In the course of 1945 and 1949, the role of the army became increasingly dominant in the Indies. The militarized approach of the Dutch to the Indonesian revolt is summarized by statistics. The Netherlands assembled 140,000 troops in the Indies by 1948. This was a tremendous effort for a country of nine million that had just seen five years of German occupation. However, the Dutch failed dramatically in raising sufficient police. There were four army personnel to every civilian police official in Indonesia by late 1948. Much of this police force of only 35,000 men had received minimal training, lacked effective leadership and proper armament, and was at times unreliable. Only 28,000 local home guards were raised for static security duties.[17] Meanwhile, the colonial administrative corps was seriously understaffed. The corps consisted of both Dutch and Indonesian governors, often in military uniform. On Java, 120 civil administrators in 1949 had to perform the job done by 230 personnel

[17] Police strength reached its height at only 35,000 at the time of the Second Police Action. Moreover, the number of home guards never exceeded 22,500. Groen, "Militant Response"; J. A. A. van Doorn and W. J. Hendrix, *Ontsporing van Geweld* (Rotterdam: Universitaire Pers, 1970), p. 140; G. C. Zijlmans, *Eindstrijd en Ondergang van de Indische Bestuursdienst* (Amsterdam, 1986), pp. 76, 88.

ten years earlier, which was already an extremely light colonial footprint.[18] In many areas, there was no civil administration to work alongside military units. In the more volatile districts, civil administrators were only given an advisory role, with the military commanders in charge of actual governance.[19] To complete the increasingly militarized approach, special military courts were put in charge of administrating justice by 1948, and the colonial intelligence apparatus was fully militarized.

By comparison, at the height of the Malayan Emergency in 1952, military forces numbered 35,000 British and Commonwealth troops. Police strength reached 28,000 that same year, which meant a ratio of almost one to one vis-à-vis the military. The number of home guards was over two hundred thousand.[20] In Malaya, a successful system for civil-military cooperation functioned on the national, provincial, and district levels. This was the so-called war-by-committee system. Within this triangular system, civil administrators played a coordinating role. Only during the height of the Emergency was a soldier, General Sir Gerald Templer, temporarily in charge of both civil and military efforts on the highest level. In contrast to the Dutch model, the British military performed its role in support of the civil power, and common law remained functional despite the state of emergency. Finally, the intelligence operation was coordinated by the civilian special branch of police.

Due to large differences in the scale of the conflict in the political, geographical, and social contexts, comparing the conflict in Indonesia to that in Malaya is risky. It is as problematic as comparing the Malayan Emergency to the American experience in Vietnam, as many counterinsurgency theorists have tried to do. Nevertheless, as with the Malaya-Vietnam comparison, the result is revealing. They offer a serious warning from the past when it comes to balancing and integrating civil and military efforts.

Conclusions

It will come as no surprise that there is no "Dutch approach" to fighting small wars and countering insurgencies in our colonial past that can explain our current extreme emphasis on the minimum use of force, on nonkinetic measures, and on the comprehensive or integrated approach. While there is continuity in our colonial past from Aceh to the War of Decolonization, current policy emphasis on other than military means is quite the opposite of past experience.

---

[18] C. Otte and G. C. Zijlmans, "Wederopbouw en Ondergang van de Indische Bestuursdienst," ZWO Jaarboek 1980, p. 182; Zijlmans, *Eindstrijd en Ondergang,* p. 57.

[19] Van Doorn en Hendrix, *Ontsporing van Geweld*, pp. 141–42.

[20] Thomas R. Mockaitis, *British Counterinsurgency, 1919–60* (New York: St. Martin's Press, 1990), p. 9; Richard L. Clutterbuck, *The Long, Long War: The Emergency in Malaya, 1948–1960* (London: Cassell, 1967), p. 43.

What explains the apparent break in Dutch responses to irregular opponents? First of all, there are two distinct lines in Dutch military and security thinking. In Europe, on the one hand, the Netherlands traditionally embraced a neutral and legalistic policy after the loss of its great power status. On the other hand, in the Indonesian archipelago, the Dutch always tended to be militaristic and realistic. This second tradition ended with our imperial retreat in 1949.

Second, when it comes to counterinsurgency experience in the twentieth century, Dutch history is characterized by discontinuity. It this sense, the Netherlands resembles the United States in various ways. Like the U.S. military—the Marine Corps in particular—the Dutch clearly had an upward learning curve from the 1890s to the Second World War. Nonetheless, both militaries emerged from the Second World War firmly believing in maneuver warfare and in decisive victory. Also, Dutch generals wanted to be Patton or Rommel. Many counterinsurgency wisdoms had been "unlearned" during the Second World War. Both the Dutch and the Americans failed to relearn, in time, how to fight an irregular opponent when faced with a Southeast Asian insurgency. Despite many tactical successes, the war in Indonesia ended in strategic defeat—an experience that will probably ring some bells in U.S. military circles. In the Netherlands and the United States, the Indonesian and Vietnamese experiences left deep wounds in the military psyche. After the frustrating fight against an illusive irregular opponent, the military establishments in both countries firmly focused on their NATO role and embraced preparations for large-scale conventional combat.

When the Cold War came to an end, however, Dutch and American experience strongly diverged. This brings us to the third explanation for the apparent change in Dutch response to insurgency. Even though the Dutch participated in many of the same peace support operations, such as in Haiti, Bosnia, and Kosovo, our appreciation of this type of military mission was radically different. U.S. policymakers and the military alike shunned "peacekeeping" as an unwelcome distraction. Dutch policymakers, and eventually also the Dutch military, embraced peace operations as their new "core business."

In short, when it comes to the use of military force in counterinsurgency operations, the Dutch have had to "upscale" for their mission in Afghanistan, whereas the Dutch, like NATO members, tend to think U.S. armed forces need to "scale down" for this type of operation. This point is, of course, open to debate, particularly as U.S. armed forces have started to embrace counterinsurgency principles and lessons and seem to apply them with some success at the tactical level in Iraq and Afghanistan. When it comes to improving civilian capabilities to work alongside military in counterinsurgency operations, the Dutch and Americans are in the same boat: a long way from where they should be. For all the talk of a comprehensive or 3-D approach, the number of Dutch civilian government officials (political advisers, development advisers, and a cultural adviser) operating alongside

1,500 military personnel in Uruzgan has varied between three and six. One can only hope that the military-civilian ratio is better in the American "whole-of-government" efforts in Afghanistan and Iraq.

PART TWO

Special Aspects of Irregular Warfare

The History of Military Commissions in the U.S. Army: From the Mexican-American War to the War on Terrorism

Frederic L. Borch

Military commissions are nothing new in the U.S. Army—or in American history. The Army first employed them in Mexico in 1847 and subsequently used them in the Civil War era, the Philippine Insurrection, World War I, and World War II. Most recently, the Army created military commissions to oversee the ongoing prosecutions of alleged terrorists held at Guantanamo Bay, Cuba.

This paper examines the history of military commissions. It begins by defining the term and examining who has the power to convene a military commission. It then looks at the structure of commissions and why they have been viewed—until now—as appropriate for prosecuting some war-related offenses. This discussion is followed by a look at military commissions from the Mexican-American War to the present War on Terrorism.

Military Commission Defined

There are four types of military tribunals: courts-martial, provost courts, courts of inquiry, and military commissions. Courts-martial, which exist in their present form under the Uniform Code of Military Justice (UCMJ) enacted by Congress in 1950, are courts for "doing justice" (although they also promote discipline). Provost courts are military-run courts in occupied territory; they exist to handle general crimes committed by civilians and were last used in occupied Germany and Japan in the aftermath of World War II.[1] Courts of inquiry, although rarely used, are also military tribunals.[2] The last type of tribunal, the military commission, is a court of

---

[1] Provost courts are convened in occupied territory for the trial of minor offenses alleged against civilians in occupied areas. During the Civil War, provost courts also tried military personnel charged with civil crimes. After martial law was declared in Hawaii during World War II—and the Territory of Hawaii civilian courts were closed—the Army operated provost courts to prosecute civilians for criminal offenses that otherwise would have been tried in civilian courts. Today, when domestic trial courts of the occupied territory are not functioning, the law enforcement arm of the occupying force (usually the provost marshal) may establish provost courts that meet the requirements of Geneva Conventions common article 3.

[2] Courts of inquiry investigate serious military incidents, such as the loss of high-value property (e.g., ships at sea) and major accidents (e.g., aircraft collisions). Courts of inquiry

extraordinarily narrow jurisdiction—it exists only during war and is designed to try enemy soldiers and unlawful combatants who violate the law of war. In this respect, it may be considered a wartime court-martial, with less restrictive rules of procedure and evidence.

Although the service originally created military commissions as temporary "field-expedient" tribunals with relatively broad jurisdiction, military commissions subsequently evolved into courts of narrow jurisdiction. By World War II, they were wartime military courts that had jurisdiction to try enemy individuals, military or civilian, who had violated the laws of armed conflict.

It follows that military commissions—because they only function during armed conflict and only have jurisdiction over war-related offenses—are of limited scope and utility. The fact that they may be used only during wartime also explains their summary nature—there is no time for a long and complicated trial during combat operations. Additionally, the relaxed standards of evidentiary admissibility and streamlined trial procedures make sense given the wartime character of military commissions—civilian-type rules are impracticable during armed conflict. For example, there is no prohibition on hearsay at military commissions because oral testimony may be difficult to obtain (witnesses present when the crime occurred may be dead, missing, or otherwise unavailable). Similarly, requiring a "chain of custody" for evidence found on the battlefield or seized from an enemy would make it difficult if not impossible to prosecute a case successfully. Consequently, military commissions generally have admitted any relevant evidence that has probative value to a reasonable person.[3]

Legal Authority to Convene Military Commissions

The president, as commander in chief, or his subordinate military commanders may convene military commissions under legal authority derived from the U.S. Constitution, Article I. In theory, any military commander may convene military commissions to try enemy soldiers or civilians who violate the laws of armed conflict. As a practical matter, however, the authority of Army commanders to convene military commissions is unsettled, as there is no written guidance or court decision on the question. It would seem reasonable, however, to assume that any commander with the authority to

---

consist of senior officers who are charged with gathering information about the incident, finding facts, and making recommendations to the commander who convened the proceedings. A recent example is the Navy's use of such a court to inquire into the facts and circumstances surrounding an American submarine's fatal collision with the Japanese research vessel *Ehime Maru* off the coast of Hawaii in 2001.

[3] In modern legal history, this "probative value to a reasonable person" language was first used in the trial of the U-boat saboteurs in 1942. When President Bush authorized military commissions for the trial of al Qaeda and other terrorists in 2001, he expressly adopted this same evidentiary standard: Military Order of 13 November 2001, "Detention, Treatment, and Trial of Certain Non-Citizens in the War Against Terrorism," sec. 4(c)(3), 66 Federal Register [F.R.] 57833-57836 (16 Nov 2001).

convene general courts-martial also would have the authority to convene military commissions.[4]

Congress, acting under its U.S. Constitution, Article II, powers, may also create military commissions. This occurred recently when Congress, in response to the U.S. Supreme Court decision of *Hamdan* v. *Rumsfeld*, enacted the Military Commissions Act (MCA) of 2006.[5] The court held in *Hamdan* that the military commission procedures established by the president in 2001 failed to satisfy international law (Geneva Conventions common article 3). The MCA is a legislative attempt to cure this deficiency.

With this as background, it follows that military commissions are very different from civilian federal courts (for example, U.S. District Courts, U.S. Courts of Appeals, and U.S. Supreme Court), which derive their authority from the U.S. Constitution, Article III, and have rules and procedures that satisfy the Bill of Rights, the Federal Rules of Evidence and Procedure, and other safeguards enacted by Congress.

Composition and Structure of Military Commissions

Given their status as war courts, military commissions historically have consisted of military officers only. Absent rules to the contrary, however, a commander convening military commissions could appoint individuals of any rank and military background to serve on the commissions. Additionally, there is no fixed number of members for a military commission: one or twelve members would be lawful.[6]

There is no requirement for legally trained counsel to sit on military commissions or to advise military commission personnel on the applicable law of armed conflict. Additionally, while there are no specific evidentiary or procedural rules for commissions, there is a general requirement that the proceedings be full and fair.

The First Military Commissions—Mexican-American War

Although George Washington's creation of a Court of Inquiry for the Revolutionary War–era trial of Maj. John Andre is often cited as the first

---

[4] During World War II, General Dwight D. Eisenhower convened military commissions in the European Theater of Operations; General Joseph T. McNarney convened them in the Mediterranean Theater of Operations; and General Douglas MacArthur convened them in the Pacific Theater of Operations.

[5] *Military Commissions Act of 2006*, Public Law 109–366, 120 Stat. 2600 (17 Oct 2006).

[6] Today, as a minimum of five members are required for general courts-martial convened under the UCMJ, it seems likely that most commanders would select a similar (or greater) number of members for their military commissions. There were, for example, five U.S. general officers on the military commission that tried Japanese General Tomoyuki Yamashita in 1945.

military commission, it was not.[7] Neither was the 1818 courts-martial and execution of two British civilians for inciting Creek Indians to wage war against the United States.[8]

Rather, the first recorded use of the military commission came during the Mexican-American War. While American units under the command of Brig. Gen. Zachary Taylor fought Mexican forces in the north, and another force of dragoons under Col. Stephen W. Kearney fought Mexican units in New Mexico and California, a third and independent force, under the command of Maj. Gen. Winfield Scott, marched on Mexico City. Leading elements of Scott's army seized the Mexican town of Tampico in February 1847, and Scott used the town as both his headquarters and a staging point for future military operations. Vera Cruz fell to the Americans on 27 March 1847, and Scott's advance to Mexico City continued until the city surrendered on 14 September.[9]

As the Americans occupied more and more Mexican territory, Scott (who had studied law at the College of William and Mary and also practiced law as a civilian) learned that he could not punish American troops at courts-martial for "common law" offenses because the then-existing Articles of War (passed by Congress in 1806) only covered military offenses like desertion and disobedience of orders. Since Congress prior to the Mexican-American War did not envisage the Army as fighting outside U.S. borders, this made sense: the Articles of War need only cover *military* misconduct because a soldier who committed murder or rape or robbery could simply be turned over to local authorities. But when U.S. regular and volunteer troops left American territory—and the reach of U.S. courts—and committed nonmilitary crimes like murder on Mexican soil, the Army could not punish them.

Scott solved this dilemma by using his powers as commander to create two new military tribunals: "military commissions" to prosecute U.S. regulars and volunteers for nonmilitary crimes and "councils of war" to prosecute Mexicans charged with war crimes. Scott saw both tribunals as temporary or interim measures, as he hoped Congress would legislate other measures for trial. In any event, on 19 February 1847, Scott published General Order 20 from his headquarters in Tampico. It read,

---

[7] General Washington convened a court of inquiry to investigate the facts of the Andre case. That court concluded (based largely on Andre's voluntary statements) that he had been engaged in a secret correspondence with General Benedict Arnold to surrender West Point. The members of the court further concluded that, while Andre was a British Army officer, he had been in disguise (and not in uniform) when he was caught in September 1780. Consequently, he should be considered a spy and put to death. Washington concurred with the finding of the court and ordered that Andre be executed.

[8] Two British nationals, Arbuthnot and Ambrister, were tried by court-martial and executed in 1818.

[9] For a concise discussion of Scott's military campaign and occupation of Mexico, see Stephen A. Carney, *The Occupation of Mexico: May 1846–July 1848*, U.S. Army Campaigns of the Mexican War (Washington, D.C.: U.S. Army Center of Military History, 2006).

> Assassination, murder, poisoning, rape . . . malicious stabbing or maiming, malicious assault and battery, robbery, theft, the wanton destruction of churches, cemeteries, or other religious edifices and fixtures, the interruption of religious ceremonies, and the destruction, except by order of a superior officer, of public or private property, whether committed by Mexicans or other civilians in Mexico against individuals in the U.S. military forces . . . should be brought to trial before military commissions.[10]

Thereafter, General Scott convened military commissions for the following offenses: manslaughter, burglary, pick-pocketing, carrying a concealed weapon, threatening the lives of soldiers, riotous conduct, and attempting to pass counterfeit money.[11] He also convened councils of war to prosecute Mexicans for law-of-war violations, with the principal offense being "guerrilla warfare" and "enticing or attempting to entice soldiers to desert the U.S. service."[12] In deciding the configuration of these military commissions and councils of war, Scott generally followed the format and procedure of courts-martial.

Together, military commissions and councils of war prosecuted 117 military commissions, 19 with Mexican defendants. They convicted Americans 68 percent of the time, and Mexicans 57 percent of the time. By the end of the war, however, Scott had eliminated the councils of war by folding them into the military commission. When Congress subsequently declined to act, the legitimacy of Scott's creation of military commissions as a legal tool came to be presumed.

Military Commissions During the Civil War and Reconstruction

During the Civil War, the Army began to convene military commissions as early as 1861.[13] During hostilities, it tried some 4,271 individuals by commission, including Maj. Henry Wirz (commandant of the infamous Andersonville prison) and the eight Lincoln assassination conspirators. The service held another 1,435 military commissions during the Reconstruction period in ex-Confederate states under military rule.

On 1 January 1862, Maj. Gen. Henry W. Hallack, commanding the Department of Missouri, published a general order that defined the nature and jurisdiction of military commissions to be used in his command.[14] Commissions convened by Hallack and other commanders during hostilities generally followed court-martial procedures, and President Abraham Lincoln personally confirmed all death sentences.

---

[10] William Winthrop, *Military Law and Precedents,* 2d ed. (Washington, D.C.: Government Printing Office, 1920), p. 832.

[11] Ibid.

[12] Ibid., p. 833.

[13] General Orders (GO) 14, 20, 118, Western Dept, 1861; GOs 24, 25, Dept of N.E. Va., 1861; GO 68, Army of the Potomac, 1861.

[14] GO 1, Dept of Missouri, 1862.

The Army employed military commissions during the Civil War to suppress guerrilla warfare, especially in the border states of Missouri, Maryland, and Kentucky.[15] For example, Robert T. Jones of Greene County, Missouri, was tried by military commission in late 1862 for "violating the law of war by letting rebels lurk in his neighborhood without reporting them to the U.S. military authorities." Greene pleaded guilty but explained that one of the "lurkers" was his brother-in-law, on whom he did not wish to inform. The commission sentenced him to six months at hard labor.[16]

Congress explicitly recognized the lawfulness of military commissions when it created the office of the Army Judge Advocate General in 1862, and President Lincoln subsequently decided that commissions were the best judicial forum at which to prosecute prominent civilian politicians who opposed the war effort, including Clement L. Vallandigham and Lambdin P. Milligan.

Vallandigham had made a speech at Mount Vernon, Ohio, in which he labeled the ongoing conflict as "a wicked, cruel, and unnecessary war . . . being waged for the purpose of crushing our liberty and erecting a despotism . . . a war for the freedom of the blacks and enslavement of the whites."[17] As General Ambrose Burnside had previously declared martial law in an area that included the state of Ohio and had also published General Orders proclaiming that those who committed acts "for the benefit of our enemies will be tried as spies or traitors," Vallandigham was swiftly brought before a military commission.

That tribunal permitted Vallandigham to have defense counsel, allowed him to cross-examine witnesses against him, and permitted him to call witnesses to testify on his own behalf. But the tribunal also convicted Vallandigham, sentenced him to be confined "during the continuance of the war," and sent him to a military prison in Cincinnati. Vallandigham then filed a writ of habeas corpus, insisting that, as he was a civilian noncombatant, the military commission had no jurisdiction over him. In 1863, the U.S. Supreme Court declined to hear his writ. In *Ex Parte Vallandigham*, it ruled that, as an Article III court, it lacked authority to review military commissions convened under Article I.[18]

After the end of the war, the Supreme Court changed course in the well-known case of *Ex Parte Mulligan*. In 1864, Mulligan had conspired with other disaffected citizens ("Copperheads") to free 8,000 Confederate prisoners held in Illinois. A military commission convened in Indianapolis found Mulligan guilty of "inciting the people to insurrection" and sentenced him to be hanged.

---

[15] Missouri alone accounted for 1,940 cases out of the total 4,271 military commissions held during the Civil War.

[16] Mark E. Neely, *The Fate of Liberty: Abraham Lincoln and Civil Liberties* (New York: Oxford University Press, 1991), pp. 171–72.

[17] *Proceedings of a Military Commission*, 21 April 1863, p. 32.

[18] *Ex Parte Vallandigham*, 1 Wallace 243 (1864).

The U.S. Supreme Court subsequently reversed Mulligan's conviction. It ruled that civilians could not be tried by military commission in any jurisdiction where the civil courts functioned. Justice David Davis, writing the majority opinion in *Mulligan*, declared that

> The Constitution of the United States is a law for rulers and people, equally in war and peace, and covers with the shield of its protection all classes of men, at all times, and under all circumstances. . . . Martial law can never exist when [or] where the [civilian] courts are open, and in the proper and unobstructed exercise of their jurisdiction.[19]

Despite the seemingly unequivocal language of *Mulligan*, military commissions continued to operate—at least in those ex-Confederate states under military rule. Congress denied legal status to those Southern state governments that refused to ratify the Fourteenth Amendment and placed them under martial law. As those states did not have functioning civilian court systems, military commissions were necessary to prevent lawlessness and disorder. But even when a state had been "reconstructed" and its civilian courts were open for business, military commissions were convened when the Army commander—who believed that Congress had given him absolute authority under martial law—decided that a military court was essential to the due administration of justice. In short, if the commander believed a local civilian court could not provide fair trials, he used a military commission. Offenses heard by these Reconstruction-era military commissions were principally crimes punished by local or common law, such as murder, manslaughter, robbery, larceny, riot, lynching, criminal conspiracy, assault, and breach of the peace. Sentences imposed by these commissions generally were the same as would have been imposed had a state court been open to hear the case.[20] The result was that the Army held more than 1,435 military commissions between April 1865 and January 1869, including the sensational trial by military commission of the eight Lincoln assassination conspirators.

Military Commissions in the Indian Wars

In 1862, Dakota Sioux warriors on the western frontier in Minnesota began killing white men, women, and children. The reasons for these massacres are complex, but as settlers increasingly encroached on Native American lands (and violated treaty obligations between the United States and the tribes), the friction resulted in bloodshed. More than 500 soldiers and settlers died before the Army defeated and killed or captured the Native American forces.

---

[19] *Ex Parte Mulligan*, 71 United States Reports (4 Wall.), 2 (1866).

[20] Winthrop, *Military Law and Precedents*, p. 853.

In 1862, Col. Henry H. Sibley, the commander of local forces, convened military commissions for the trial of the Sioux captives for the earlier massacres. They had minimal rules of procedure and little regard for the rules of evidence. As President Lincoln observed in reviewing the 303 death sentences imposed by the Dakota commissions, "The trials had become shorter and shorter as they progressed."[21] Lincoln commuted the sentences or pardoned most of the defendants; he authorized a total of forty executions. Thirty-eight were hanged on 26 December 1862.[22]

Military Commissions During the Philippine Insurrection

Military commissions were used during the fighting in the Philippines after the Spanish-American War. General Arthur MacArthur (father of Douglas) declared martial law "and relied on a mix of military commissions and Army provost courts to discipline the local population."[23] Military commissions generally followed the rules of evidence and procedure used for courts-martial under the Articles of War. They also recognized combatant immunity under the law of war, which meant that some insurgents had a defense to murder.

Military Commissions During World War I

Apparently the only military commission to arise during World War I involved a German spy, Lothan Witzke.[24] He entered the United States from Mexico using a Russian passport (under the alias Pablo Waberski) and with a mission to carry out sabotage. He was arrested in Nogales, New Mexico, and tried by a military commission consisting of two Army brigadier generals and three colonels. Witzke was prosecuted for violating Article 82, Articles of War. This provision made it a crime for "any person in time of war to . . . lurk or act as a spy in or about any of the fortifications, posts, quarters of encampments of the armies of the United States." The commission found him guilty and sentenced him to death; President Woodrow Wilson commuted his sentence to confinement

---

[21] Peter J. Richards, *Extraordinary Justice: Military Tribunals in Historical and International Context* (New York: New York University Press, 2007), p. 29.

[22] At the time, the Lincoln administration was under considerable military pressure (victory at Gettysburg was still more than six months away), and some believe Lincoln took a hard line with Native American warriors to forestall further unrest.

[23] Louis Fisher, *Military Tribunals and Presidential Power: American Revolution to the War on Terrorism* (Lawrence: University Press of Kansas, 2005), p. 80.

[24] The United States conducted no other military commissions. The French, however, prosecuted about 1,200 Germans at military commissions; the Belgians about 80 more. Some Germans also were tried by the Supreme Court of Leipzig, but the results were so unsatisfactory that the Allies learned that it was foolish to permit a defeated enemy to try his own suspected war criminals.

at hard labor for life. After the Army judge advocate general recommended his release, however, Witzke was deported to Germany in 1923.[25]

Military Commissions in World War II

From 1942 until 1949, the Army prosecuted military commissions in various locations, including China, France, Germany, Italy, Japan, the Philippines, and the United States.

President Franklin D. Roosevelt convened the first military commission in 1942. The case arose out of the capture of eight German saboteurs on U.S. soil—four on Long Island, New York, and four in Jacksonville, Florida. The public demanded immediate execution of the Germans. Roosevelt, however, decided that the circumstances required some type of judicial proceeding. He preferred a court-martial, but then Attorney General Francis Biddle argued instead for a military commission, chiefly because there would be relaxed rules of evidence and, arguably, no right to appellate review.

The Supreme Court's ruling in the case, *Ex Parte Quirin,* written after the execution of most of the saboteurs, found that they did have a right to judicial review. The *Quirin* decision, however, while establishing civilian judicial review of military commissions, also concluded that the instant proceedings had been lawful—thus making the decision politically acceptable.[26]

Starting in Europe in 1944 (with the trial by military commission of German soldiers caught wearing Army uniforms during the Battle of the Bulge), the Army prosecuted enemy soldiers and civilians for a variety of war-related offenses. For example, it tried and convicted German civilian farmers for murdering downed Army Air Forces pilots and aircrew. In other cases, the Army tried, convicted, and executed soldiers and civilians for murdering Jews and other inmates at concentration camps in Belsen, Dachau, and Flossenburg. In Italy, German soldiers (including one general officer) were prosecuted before commissions for ordering the execution of U.S. prisoners of war in Italy.[27]

In the Pacific, after the end of hostilities in 1945, the Army successfully prosecuted Japanese soldiers for murdering prisoners of war and committing other atrocities upon civilians. The best known was the trial by military

---

[25] Ibid., pp. 87–89.

[26] In 1944, two more German spies were caught in New York and tried by military commission (but using courts-martial procedures). While they were convicted and sentenced to death, both were spared after Roosevelt died and war with Germany ended.

[27] For example, German General Anton Dostler was tried at a military commission in Rome, Italy, 8–12 October 1945. Charged with ordering the execution of two officers and thirteen enlisted men from an American reconnaissance battalion that had been captured while on a sabotage mission, Dostler claimed that he had acted in obedience to orders from Hitler to execute all commandos. The commission rejected the defense and found Dostler guilty. He was executed by firing squad—the only German general officer executed on the sole authority of the United States. UN War Crimes Commission, *Law Reports of Trial of War Criminals*, I:30 (London, 1947); Richards, *Extraordinary Justice*, pp. 117–18.

commission of General Tomoyuki Yamashita. Brought to trial in October 1946 on the charge that he had failed to discharge his duty to restrain his troops in the Philippines from murdering, raping, and otherwise brutalizing thousands of civilians, Yamashita was found guilty and executed.[28]

The service conducted 950 military commissions during and after World War II: 491 in Europe and 459 in the Pacific.[29] About three thousand German and Japanese defendants were tried, and no serious scholar disputes that the trials generally were fair (witness their 10 percent acquittal rate).[30]

Military Commissions in the War on Terrorism

In the aftermath of al Qaeda's 11 September 2001 attack on the World Trade Center and the Pentagon, President George W. Bush announced the establishment of military commissions to try al Qaeda members and others for acts of terrorism against the United States.

The military commissions created in Bush's Military Order 1 were extremely narrow in jurisdiction—both in terms of personnel subject to trial and the kind of offenses triable. For example, they applied only to non-U.S. citizens and only for terrorist-related offenses that "threaten to cause, or have as their aim to cause, injury to . . . the United States."[31]

While the order required "full and fair" proceedings, it allowed classified trials, closed to the public, to avoid the disclosure of "state secrets," and trials *in absentia*. The order also permitted minimal rules of evidence: any evidence with probative value to a reasonable person was admissible; there were no restrictions on hearsay; and evidence obtained through torture or coercive interrogation techniques was admissible.

Critics immediately complained that the Guantanamo military commissions violated the Geneva Convention Relating to the Treatment of Prisoners of War. They argued that Article 102, which requires that any prisoner of war be sentenced by the same courts using the same standards of justice enjoyed by U.S. soldiers, meant that military commissions had to comply with current

---

[28] While the U.S. Supreme Court subsequently affirmed the proceedings in *In re Application of Yamashita*, 327 United States Reports 1 (1946), the military commission proceedings have been widely criticized by historians and do not meet today's standards of legal due process.

[29] These numbers include military commissions conducted by the Navy on Guam.

[30] As noted by John A. Appleman, *Military Tribunals and International Crimes* (Indianapolis, Ind.: Bobbs-Merrill, 1954). Most Allied nations, appreciating that American military commissions were a useful tool, convened their own commissions to prosecute German and Japanese citizens who had violated the law of war. The Dutch conducted commissions in the Netherlands East Indies; the Chinese prosecuted Japanese soldiers at commissions in China; and the French conducted commissions in both France and Germany. The Soviet Union also held thousands of military commissions, but information and numbers on these proceedings have never been released.

[31] Military Commission Order (MCO), *Detention, Treatment, and Trial of Certain Non-Citizens in the War Against Terrorism*, 13 November 2001, 66 F.R. 57834, sec. 2(a)(1)(ii).

courts-martial practice and procedure under the UCMJ. The government responded that al Qaeda members and other detainees at Guantanamo were not prisoners of war and, consequently, Article 102 did not apply.

In any event, on 29 June 2006, the U.S. Supreme Court limited the authority of the president to hold the Guantanamo commissions. While the court did not decide the applicability of Article 102, it ruled in *Hamdan* v. *Rumsfeld* that the tribunals created under the president's executive order failed to comply with Geneva Conventions common article 3.

This provision, which is "common" to all four Conventions, prohibits "the passing of sentences and the carrying out of executions without previous judgment pronounced by a regularly constituted court, affording all the judicial guarantees which are recognized as indispensable by civilized peoples." According to the U.S. Supreme Court, the Bush administration's commissions failed to satisfy this standard.

In December 2006, Congress exercised its power under Article II and created a military commission system under the MCA of 2006. The act affirms that commissions are military tribunals of extremely narrow jurisdiction and establishes "procedures governing the use of military commissions to try alien unlawful enemy combatants engaged in hostilities against the United States for violations of the law of war." The statute expressly states that the military commissions comport with the requirements of the Geneva Conventions common article 3.

The MCA states that procedures for military commissions are modeled on the procedures for general courts-martial under the UCMJ, except that there is no right to speedy trial, no right to a pretrial investigation, and no right to be warned against self-incrimination. Additionally, MCA procedures permit the admissibility of hearsay evidence, as well as evidence obtained through coercive interrogation techniques (and inhumane and degrading treatment). Whether the MCA will satisfy the Supreme Court almost certainly will be litigated in the near future.

In the meantime, the first case under the new system has been completed. In March 2007, Australian David Hicks pleaded guilty in a negotiated agreement to the charge of "providing material support for terrorism." He was sentenced to serve nine months—which he will serve in Australia.[32]

Conclusion

Military commissions originated during the Mexican-American War as a gap-filler to prosecute U.S. regular and volunteer troops for common law offenses not covered by the 1806 Articles of War. The commissions—along with councils of war—also tried Mexicans for violating the laws of war. By

[32] Hicks was captured in Afghanistan and spent more than five years in detention at Guantanamo Bay, Cuba. For more on Hicks, see Leigh Sales, *Detainee 002: The Case of David Hicks* (Melbourne, Australia: Melbourne University Press, 2007).

the end of the American occupation of Mexico in 1848, military commissions had emerged as legitimate judicial tools.

By the end of the nineteenth century, the jurisdiction of military commissions had narrowed significantly. Congress had amended the Articles of War to bring common law offenses under the jurisdiction of courts-martial, which, as a result, restricted military commissions to trying violations of the law of war.

The zenith of the military commission was World War II. From 1942 to 1949, military commissions tried hundreds of German, Italian, Japanese, and other enemy soldiers and civilian defendants for violations of the laws of armed conflict.

Today, more than enough historical precedent exists to permit the United States—or any nation state—to use military commissions to prosecute "terrorists," provided there is an armed conflict. Consequently, military commissions could lawfully be used to prosecute al Qaeda terrorists or Iraqi or Taliban insurgents attacking U.S. forces today in Afghanistan and Iraq. But those military commissions must comply with Geneva Conventions common article 3 to be lawful (and perhaps Geneva Prisoners Convention Article 102). The MCA of 2006 attempted to meet the minimum due process standards required by the Supreme Court in *Hamdan* v. *Rumsfeld*, but whether this legislation is sufficient remains problematic. The end result is that the future of military commissions in the Army is very much in doubt.

Intimidation, Provocation, Conspiracy, and Intrigue: The Militias of Kentucky, 1859–1861

John A. Boyd

Militias are paramilitary organizations that have made headlines from Yugoslavia to Somalia to Iraq. Militias brought death to untold thousands in Rwanda and now in Darfur. In Iraq, few have not heard of the Shiite Badr militia or Moqtada al-Sadr's militia of the downtrodden masses, the Jaysh al-Mahdi. Militias today invoke a certain sense of fear and dread, as well they should. To modern Americans, they now symbolize death squads, fanaticism, anarchy, and destruction. However, for better or ill, loath them or embrace them, militias have their uses.

During the secession crisis of 1861 prior to the outbreak of the American Civil War, the militias of Kentucky—some pro-Union, pro-southern, and even pro-neutral—played a pivotal role in determining whether the Bluegrass state would stay out of civil war or enter the conflict on the Union or Confederate side. Implausible as it may seem, they did this without firing a shot.

The militias of Kentucky cannot be understood properly without understanding the martial heritage of antebellum America in general and of Kentucky in particular. In 1860, schoolbook histories immortalized a republic born in blood, dwelling on the rattle of musket and the clash of bayonet in the founding of free institutions. A war record could turn a backwoods politician into a president, and, at least in the South, the readiness to use violence to vindicate one's honor actually improved many a statesman's standing. What was true elsewhere held truer still in the "dark and bloody ground" of Kentucky. Statues in Daniel Boone's honor showed him, quite uncharacteristically, as an Indian-fighter, and legends about him coated him in the glamour of a bloodlust utterly foreign to the man himself. Kentucky still had its veterans of the War of 1812, and every town could still point to its old-timers raised in the days of the Indian wars. If backwoodsmen looked on the landed gentry with suspicion, one reason was their suspected lack of fighting qualities. Even Whig and later Unionist Senator John Jordan Crittenden found advantages in letting voters know of his accomplishments during the War of 1812, for as one contemporary declared, over the next half-century "to have fought at the Thames was the 'open sesame' to public and political honor." Mexican War veterans won the same acclaim if not notoriety in 1850s politics.[1]

---

[1] Quote from Albert Kirwan, *John J. Crittenden* (Lexington: University of Kentucky Press, 1962), p. 25, and see also p. 26. See also John Hope Franklin, *The Militant South, 1800–1861*

In peacetime, with the Indian wars receding into the distant past, Kentucky could boast of several well-trained prewar militia companies, among them John Hunt Morgan's Lexington Rifles and Simon B. Buckner's Louisville Citizens' Guard. No patriotic celebration was complete without a turnout of the local militia. Spectators could watch close-order drill, rifle volleys, and mock battles, and, from the size of the crowds that turned out, thought it very good theater. Throughout America, "in the everyday life of the city, private military clubs ranked first among the street performers." The public considered Morgan's Lexington Rifles the best and demanded its attendance at holiday occasions like Washington's Birthday or Independence Day. Sometimes the Rifles' best performances at resorts like Paroquet Springs and Crab Orchard lasted a full week.[2]

That meant less and more than it might seem. The prewar militia, as Morgan's own contingent showed, was primarily a social organization. Young men were eager to join. Membership gave them a chance to parade in splendid uniforms and perform elaborate maneuvers with sabers and rifles in front of a vast audience, including eligible young ladies. Militia companies also sponsored charities and dances. They staged shows for worthy civic causes. And finally, to their members, they offered all the benefits of a fraternal society.[3]

Militia membership gave many a young man a sense of belonging. Most military companies had fewer than fifty members. A company was built on a common culture, shared interests, and a general sense of brotherhood. It encouraged political fealty and social cohesion, especially when, as often happened, the men elected their own officers or were recruited by the man who paid the organization's bills. The loyalty of Morgan's men was well known throughout Kentucky long before the war and, with it, their slogan, "Our laws, the commands of our Captain."[4]

Most active militia companies in the 1850s were inclusive by being exclusive. Their sense of belonging rested on being separate from those outside. "This little company of citizen soldiers were in their conceit and imagination very important and consequential fellows," an ex-lieutenant later wrote scornfully of the Flat Rock Greys. "Invited to all the noted gatherings and public affairs of the day, dressed in gaudy and flashy uniforms and flying plumes, filled with pride and *conceit*." For obvious pocketbook reasons, poorer men rarely joined such units. Panoply did not come at cut rates. The

(Cambridge, Mass., Harvard University Press, 1956). This paper is a slightly expanded version of the one read at the conference of Army historians held in Arlington, Virginia, in August 2007.

[2] Susan G. Davis, *Parades and Power: Street Theatre in Nineteenth-Century Philadelphia* (Philadelphia: Temple University, 1986), pp. 49–53; James A. Ramage, *Rebel Raider: The Life of General John Hunt Morgan* (Lexington: University of Kentucky Press, 1986), pp. 35, 41–42.

[3] F. C. Harrington, *Military History of Kentucky* (Frankfort, Ky.: Military Department of Kentucky, 1939), p. 141.

[4] William Henry Perrin, ed., *History of Fayette County* (Chicago: O. L. Baskin, 1882), p. 444.

Lexington Rifles paraded in bearskin grenadier-style hats, and its duty caps had the seal of the state of Kentucky; another company used the tricorner hat of the American Revolution. Typical styles of the day included elaborate tunics, buttons, ribbons, buckles, and belts.[5]

The fraternal benefits of militia membership, then, gave members psychic rewards, but they were far from the universal service—the kind of citizen armies that Americans liked to think would save them from foreign foes in wartime. They were more for play than work because there was not much work for them to do. Theoretically, they stood ready to keep the civil peace and maintain order. A few actually served that purpose. Lawyers outraged at an outbreak of vigilantism in Louisville in 1857 formed Buckner's Citizen Guards. At one point, Louisville's fire chief led a company of militia that stood prepared to help on-duty firefighters if an emergency arose. (Curiously, there is no record of any militia company created in the 1850s specifically to forestall slave revolts.) But all these were rarities.[6]

Conceivably, Kentucky could have created a state militia to go along with all these private companies, one open to all citizens. Such a system had existed once. But with no Indian menace requiring a citizen army, it seemed an anachronism. The state legislature allowed the militia system to become dormant in 1854. The commonwealth dropped requirements for regular militia musters. It had passed out weapons but lost track of where they had gone. Those arms remaining were outdated muskets, usually in such poor repair as to be practically worthless. "There is in fact, no organized militia in the State," a governor summed up in 1856.[7]

The Creation of the Kentucky State Guard

The raid on Harpers Ferry, Virginia, led by John Brown in 1859 changed all that. Fears that abolitionists might cross into Kentucky spreading mischief, if not murder, went far back into the state's past. Now that a group of abolitionists had attacked Virginia, a sister border slave state, an attack on Kentucky would inevitably follow, or so most men believed. It seemed obvious to Governor Beriah Magoffin. Kentuckians had no way of knowing "at what moment we may have need of an active, ardent, reliable, patriotic, well-disciplined, and thoroughly organized militia," he informed the legislature in December 1859. If "some of the most distinguished leaders and ministers of the Abolition and Republican party" did not plan Brown's invasion, they surely knew of

---

[5] Quote from L. D. Young, *Reminiscences of a Soldier of the Orphan Brigade* (Paris, Ky., 1918), p. 12, and see also p. 11. *Lexington Observer and Reporter*, 16 Sep 1857.

[6] Prior to John Brown's Raid, Kentuckians had little reason to fear slave revolts—there had been none. After the raid, they effectively expelled abolitionist John G. Fee of Berea and others without incident.

[7] Richard G. Stone, *A Brittle Sword: The Kentucky Militia, 1776–1912* (Lexington: University of Kentucky Press, 1977), p. 59.

it, approved it, and helped it out. Of course, Magoffin erred. No Republican leader was involved, and the event shocked even radicals like Pennsylvania Congressman Thaddeus Stevens, who remarked, "You hung them exactly right, Sir," to a Virginia representative after Brown's execution. But Magoffin had made a convincing point to quite receptive lawmakers.[8]

In planning for a revived militia, the governor turned to 37-year-old Simon B. Buckner, a West Point graduate and Mexican War veteran living in Louisville. Buckner quickly submitted a proposal for reorganization so detailed that it even prescribed how many ostrich plumes the governor ought to wear, not to mention color. The legislature gave swift approval, bringing a new militia—the Kentucky State Guard (KSG)—into existence.[9]

To command it, Magoffin appointed Buckner himself to the office of state inspector general with the rank of militia major general. The appointment gave Buckner considerable powers and responsibilities. He could activate the militia in any emergency and for an indeterminate period.

Buckner had the energy and enthusiasm for the task. At once he set to work scouring state records for weaponry to arm his forces. By early 1861, he could report that the state owned 11,283 muskets, 3,159 rifles, 2,873 cavalry arms outfits, and 53 field pieces—more weapons than the state of Ohio controlled at the onset of the Civil War. Around him he gathered a talented staff, among them Abraham Lincoln's brother-in-law, Ben Hardin Helm, a U.S. Military Academy graduate who became assistant inspector general. He chose surgeons, commissary officers, quartermasters, and even chaplains with an eye to both their military capacity and the political ramifications. The choices proved effective ones.[10]

Buckner found much of his army ready-made. Existing companies quickly joined the KSG, with Morgan's Lexington Rifles among the first officially mustered in. Other militia units, organized in reaction to John Brown's Raid, also joined up. By August 1860, Buckner oversaw a force of forty-nine militia companies—some 2,500 men. That month, he put them on display, ordering a week-long training encampment near Louisville. It was a highly publicized event. Seeking a tactically proficient force, the inspector general reserved the first three days of training at the newly christened Camp Boone for officers only.[11] One could, perhaps, see it as a rebel army in embryo, but the striking thing about the occasion was how much the KSG fit the social character and

[8] *Journal of the House of Representatives of the Commonwealth of Kentucky . . . 1859* (Frankfort, Ky., 1859), p. 40; Fawn M. Brodie, *Thaddeus Stevens: Scourge of the South* (New York: W. W. Norton, 1959), pp. 133–34.

[9] An Act for the Better Organization of the Kentucky Militia, approved 5 March 1860, in *Acts of the General Assembly of the Commonwealth of Kentucky Passed at the Session Which Was Begun and Held in the City of Frankfort on Monday, the Fifth Day of December, 1859 and Ended on Monday, the Fifth Day of March, 1860* (Frankfort, Ky., 1860), pp. 142–71.

[10] Harrington, *Military History of Kentucky*, p. 147; Stone, *Brittle Sword*, p. 62.

[11] The camp was set up at the South West Agricultural and Mechanical Association, of which secessionists Blanton Duncan and Thomas H. Hunt were members. See *Sketches of Camp*

behavior of militia companies from times past. Alcohol flowed freely. One photograph of some officers displays a whiskey bottle carelessly left in the foreground. Soldiers made money too: one Sunday, 3,500 guests paid a 25-cent fee to watch a mock battle.[12]

Yet, upon closer inspection, the KSG had gone beyond people playing soldier—the politics of Union or secession had intruded. For some men, among them visiting secessionist Blanton Duncan and Maj. Thomas H. Hunt, commander of Louisville's KSG regiment, the encampment gave them their first ever opportunity to exchange views and forge secret alliances with militia leaders from all over the state. With the 1860 presidential campaign in full swing, it is reasonable to speculate that, around campfires and over drinks, talk may well have turned to politics and what Kentucky would do if worst came to worst. The induction of the governor and many KSG officers into the "ske-tie-tu-rus" society (code for state rights) may have been as much a burlesque as it seemed, but what did it mean when select officers became members of the Knights of the Golden Spur? Was this mysterious order a thinly disguised surrogate of the Knights of the Golden Circle, an organization dedicated to the conquest and creation of an American empire for slavery?[13] No hard evidence has survived, but, quite possibly, that August 1860 encampment served as a school for political education for those willing to listen.[14]

Certainly Buckner created a military force loyal to himself and ready to follow his commands—one with a professional officer corps that placed personal loyalty above political disagreement. Officers like Helm, Hunt, Morgan, Lloyd Tilghman, and Thomas L. Crittenden (the senator's son) owed first allegiance to Buckner, whatever their own views of the rights and wrongs of the sectional conflict. That loyalty mattered; without it, Buckner's leadership during the secession crisis of 1861 would not have been as consequential as it would prove it to be.

Just as important, most Guard officers plainly held to the southern rights position. That did not make them disloyal to the Union in 1860. Southern sympathies, fealty to Kentucky, and allegiance to Buckner were perfectly compatible with love for the Union at that time. But when sectional and national loyalties began to pull men apart, Buckner's influence provided one of the strongest forces to hold them together and to keep men of Union and secessionist sympathies working together for the longest possible time.

Boone: The First Encampment of the Kentucky State Guard (Louisville, Ky.: G. T. Shaw, 1860), pp. 15–20; *Louisville Daily Courier*, 27 Aug 1860.

[12] *Louisville Daily Courier*, 28 Aug 1860; *Sketches of Camp Boone*, p. 25.

[13] The Knights of the Golden Circle, led by founder George W. L. Bickley, constituted a secret military society dedicated to the expansion of slavery throughout Latin America. The primary goal of the organization was to annex northern Mexico and create a new slave state.

[14] Ltr, Blanton Duncan to Stephen A. Douglas, 7 Mar 1861, Douglas MSS, University of Chicago; Basil W. Duke, *History of Morgan's Cavalry* (Cincinnati, Ohio: Miami Printing and Publishing Co., 1867), p. 36.

Finally, an effective KSG just may have given Governor Magoffin more confidence in taking the political positions he did. The Kentucky Constitution of 1850 had made the office of governor a near-figurehead and had drastically trimmed the governor's powers of patronage. But no provisions tampered with the governor's powers as commander in chief. Now Magoffin had something worth commanding: an instrument of potentially great power; an army eventually numbering 4,000 men that conceivably might seize the Bluegrass state in a secessionist coup.

The Secession Crisis, 1861

Lincoln's election and the secession of South Carolina shattered and realigned the political parties of Kentucky beyond recognition. By mid-January 1861, Kentucky's Democratic Party split into pro-Union (Douglas Democrats) and pro-southern (Breckinridge Democrats) factions. Meeting in private, leaders of the old pro-Union Whig Party and pro-Union Democrats joined forces to form a Union party officially named the Union-Democracy (UD). Pro-southern Democrats countered this Union party realignment several months later, creating the Southern Rights Party (SRP). Locked in a political struggle to determine Kentucky's allegiance to North or South, the UD and the SRP competed feverishly for the hearts and minds of Kentuckians.

These political maneuverings and machinations left the commonwealth's Governor Magoffin almost as a man without a country—he had lost his formerly unified Democratic Party and, with it, his legislative majority. While Magoffin publicly espoused southern rights, he favored secession in his heart. But he understood better than most that loyalties in his state divided evenly and that his beloved commonwealth could rapidly disintegrate into anarchy and chaos. He feared Kentucky—a border slave state—would be ripped apart and destroyed due to its geographic proximity to both sides, as well as consumed in an internecine civil war.

And so Magoffin sat on the fence. He attempted to ride the crisis out, hoping for a sign and waiting for some indication of which side Kentucky should take. After all, many pundits predicted a short, ninety-day war. He had everything to gain and nothing to lose by waiting it out. He resisted Lincoln's calls for troops after Fort Sumter had been fired upon, saying Kentucky would not supply soldiers for the "wicked purpose of subduing her sister southern states," but at the same time he spurned southern commissioners and troop requests from the newly formed Confederate States.

The aftermath of Fort Sumter tested the loyalties of the newly minted KSG militia. Secessionist Blanton Duncan and other radicals had perfected a scheme to muster rebel troops. Duncan recruited a regiment of Kentuckians for immediate Confederate service and, before the firing on Fort Sumter, had arranged that the rebels gather in Louisville for transfer south whenever Duncan gave the word. "I have tendered to Genl Davis a regiment of 1,000 men *well drilled* and *prepared to march at a moment's notice*," Duncan misinformed Senator Stephen A. Douglas in March 1861. "Of course you will keep this *private*."[15]

Acting on Duncan's plan after Sumter, the first rebel volunteers started arriving in Louisville on 12 April. Local authorities worried and Unionists openly charged that their real aim was to take over the city. Word soon spread that Buckner himself had arranged with Duncan to keep his soldiers home a little longer, "in the event that their services may be needed for the defense of Kentucky from Northern aggressions." "Our city is assuming a decidedly military aspect," one Louisville man noticed. "The tread of armed men is heard in our streets every day and night."[16]

Alarmed, Louisville Mayor Thomas Crawford had earlier asked the governor to place a KSG company at his disposal in the event of trouble. Buckner now obliged, ordering Major Hunt to have a company report to the newly elected mayor, John M. Delph. On 18 April, Buckner detailed the Citizen Guards as a special police and ordered them to guard the city battery throughout the night. All that week, KSG companies shared the task of guarding the city. When a hundred-plus Confederates under Joseph Desha arrived from Cynthiana, Delph ordered state guardsmen "to be in their armories ready at a moment's notice," for "trouble might ensue." But calm prevailed. When Duncan's Confederate regiment, now four hundred strong, marched to the Louisville and Nashville railroad station for the journey south, they furled their banners, except for a Captain Harvey "who flung his to the breeze." The city fathers must have breathed a sigh of relief, and they were right to do so. The departure of Duncan's Confederates erased the most serious threat to the internal peace of Kentucky for the moment, and the KSG had proved loyal to the commonwealth in its first test.[17]

The Union Home Guard

If Buckner and his cohorts did not see the issue clearly at first, Unionists of the new Union-Democracy Party did. In order to save Kentucky for the Union, they must defeat or disarm the pro-southern KSG. The belief in a secessionist plot to seize the state was fixated in the minds of Union-loving men. According to the *Louisville Journal*, they saw daily indications that "the secessionists of Kentucky are moving in a secret conspiracy to take the State out of the Union by a sudden, violent and if necessary, bloody process." Unionists needed military force to guard against this, and it could not, due to

---

[15] Ltr, Duncan to Douglas, 7 Mar 1861, Douglas MSS, University of Chicago.

[16] *Louisville Daily Courier*, 13, 18, 20 Apr 1861.

[17] Kentucky State Guard Special Order 56, Personnel Records, Kentucky National Guard, Frankfort (cited hereafter as KSG Special Orders); Alfred Pirtle, Journal, 20, 22 Apr 1861, Papers of Alfred Pirtle, Filson Historical Society, Louisville, Ky.; *Louisville Daily Courier*, 20 Apr 1861.

political sensitivities, be a force of federal soldiers. Consequently, the Unionists created local Union Home Guard militias.[18]

The presence of well-trained, active KSG militia companies (composed of southern rights members) could and did intimidate Union men. Units numbering forty to fifty men with muskets and bayonets, chanting secession slogans, had the ability to frighten average citizens who had never seen armed formations during their lifetime. To civilians who had never seen large numbers of troops, uniformed men with rifles in their hands made a powerful impression. Mexican War veterans—men such as Buckner, Lovell Rousseau, or Morgan—knew better. They understood that the KSG and its tiny battalions could not effectively seize and hold the commonwealth. But to the average Kentuckian, a militia company of forty armed men seemed a mighty host. One Unionist complained that the KSG was "daily becoming insolent and overbearing and disposed to violence." Another attacked the KSG, saying, "the thing they most respect is the strong arm with a weapon at the end of it."[19]

Unionist Garrett Davis was convinced that the military situation in Kentucky was at a flash point. "The Union men of Kentucky express a firm determination to fight it out," reported his contact, Union General George McClellan. "Yesterday Garrett Davis told me, 'We will remain in the Union by voting if we can, by fighting if we must, and if we cannot hold our own, we will call on the General Government to aid us.' . . . [He] convinced me that the majority were in danger of being overpowered by a better-armed minority."[20]

To counter the Southern Rights Party, the Union-Democracy immediately called for the creation of local Union Home Guard militias to protect hearth and home. These independent companies—in reality, political militias—were organized and equipped by prominent Union men in Kentucky's major cities and towns. On 25 April, Louisville opted to recruit two regiments of "police" and designated the new, pro-Union mayor John Delph as the commander in chief. "We are in favor of the Home Guards," trumpeted the *Yeoman*, "and in favor of distributing arms judiciously among them, for local defense of the counties."[21]

The elderly John J. Crittenden, a veteran of the War of 1812, captured headlines when he announced his membership in the 162-member Frankfort Home Guard. One can only imagine the impression it made upon Kentuckians seeing the ancient Crittenden, rifle in his hands, but the message was manly

---

[18] *Louisville Journal*, 8 Jun 1861.

[19] Ibid., 2 Jul 1861

[20] *War of the Rebellion: A Compilation of the Official Records of the Union and Confederate Armies*, 128 vols. (Washington, D.C., 1880–1901), ser. 3, vol. 1, p. 236; Ltr, McClellan to Townsend, 17 May 1861.

[21] *Frankfort Yeoman*, 16 May 1861.

and clear: the Union men of Kentucky would fight, and the Bluegrass state would go down in blood should secession be attempted.[22]

Covert Operations: The "Lincoln Guns . . . Neutrality with a Vengeance"[23]

Like their KSG counterparts, the chief problem for Unionists was finding weapons with which to arm themselves. Fortunately for Kentucky Unionists, a covert operation under the direction of Navy Lt. William "Bull" Nelson came to their rescue. Nelson, stationed at Washington, D.C., met secretly with President Lincoln and proposed smuggling guns to Union men. Using Lincoln's close personal friend, Joshua F. Speed, as his point of contact in Kentucky, Nelson met secretly with key Union leaders James Harlan, Charles A. Wickliffe, Garrett Davis, Thornton F. Marshall, and John J. Crittenden in Frankfort on or about 6 May. They founded the Union Defense Committee. All were of the "profound conviction that the guns were necessary to the salvation of the state."[24] And guns, later called Lincoln Guns, they would get.

Nelson signed for his first consignment of Lincoln Guns (5,000) at Cincinnati on 5 May, just days after his conference with Lincoln. He then shipped part of the consignment to Jeffersonville, Indiana, where 1,200 rifles were quietly issued by his agents to the Louisville Home Guard. Following this, Nelson put part of his Cincinnati cache on board Kentucky Central trains (17 May) shipping them to Paris and Lexington, saturating the Bluegrass counties. Having exhausted his initial supply by 5 June and promised an additional 5,000 rifles by Lincoln, Nelson continued his weapons operation from Cincinnati. In all, Bull Nelson oversaw the distribution of 23,000 rifles in Kentucky.[25]

Once the smuggled arms were in the hands of Union Home Guard men, it was never intended that they remain a secret. Instead, the news was leaked with great fanfare and effect. Southern rights leaders protested that the Lincoln Guns were designed to "begin civil war in Kentucky." Under headlines entitled "THE CONSPIRACY," the SRP accused the UD of duplicity, crying, "Companies of home guards . . . have driven every Southern man from their ranks." The SRP also exaggerated the number of weapons, severely damaging its own cause; at

[22] Ibid., 30 Apr 1861. The willingness of Crittenden and other Union elders to shoulder a musket for their cause conjures up the image of Iranian revolutionary Mullahs circa 1979, who were no less determined.

[23] Ltr, R. H. Stevenson to T. B. Stevenson, 18 May 1861, T. B. Stevenson MSS, Cincinnati Historical Society Library.

[24] Kirwan, *Crittenden*, p. 436. Quote from Daniel Stevenson, "General Nelson, Kentucky, and Lincoln Guns," *Magazine of American History* 10 (1883): 122, and see also p. 121.

[25] Stevenson, "Nelson, Kentucky, and Guns," pp. 119, 123–25. Most of the July consignments (13,000 rifles) were stored for issue to Tennessee units, which would be recruited in August.

one point, it overestimated the 2,500 rifles as 15,000. A Unionist later quipped, "Each gun was thus made to have the moral effect of three or four."[26]

The psychological effect of militia weaponry had an important effect upon friend and foe. It tipped the balance in favor of the Union. Watching men parading down Main Street (Danville) with their new Lincoln Guns, Speed S. Fry was amazed at public reaction. "It would be impossible," Fry observed, "for any one to describe, in language sufficiently strong, the consternation expressed in the countenances of these people, when they beheld my company of a hundred men file down Main street, with bayonets glistening in the sunlight, pointed above their heads, and nodding to and fro as they 'kept step to the music of the Union.'" Guns, in the opinion of most Union men, "had a wonderfully quieting effect in the communities into which they were introduced."[27]

A Neutral Regime?

With KSG and Home Guard companies threatening and taunting each other, Governor Magoffin—sometimes labeled "His Hesitancy"—worried that Kentucky teetered dangerously on the brink of destruction. In one of the most extraordinary actions of the Civil War, he proclaimed the Commonwealth of Kentucky neutral on 20 May 1861. In his proclamation, the governor urged Kentuckians "to refrain from all words and acts likely to engender hot blood and provoke collision."[28] He failed to mention his intention to mobilize units of the KSG to enforce neutrality throughout the commonwealth.

The recent crisis in Missouri—a bloody day in St. Louis on 10 May that sparked internal civil war in that state—plus public knowledge of Nelson's smuggled Lincoln Guns resulted in special KSG military orders.[29] Buckner, with Magoffin's concurrence, determined to field a pro-neutral thousand-man militia army. In addition, Buckner sent orders to the Lexington Battalion (Roger W. Hanson commanding) to activate a camp of instruction on 20 May (the same day as Magoffin's proclamation). Rumor had it that Lincoln's troops would attack the KSG on 21 May. Was it all just coincidence? That same day, the Louisville KSG battalion was ordered by KSG headquarters (Louisville)

[26] First quote from *Louisville Daily Courier*, 20 May 1861. Second quote from Stevenson, "Nelson, Kentucky, and Guns," p. 126, and see also pp. 122–23.

[27] Thomas Speed, *The Union Cause in Kentucky, 1860–1865* (New York: G. P. Putnam's Sons, 1907), p. 109; Stevenson, "Nelson, Kentucky, and Guns," p. 131.

[28] Speed, *The Union Cause in Kentucky*, pp. 47–49. The legislature had also passed a neutrality resolution on 16 May.

[29] On 10 May 1861, the Missouri State Guard was surrounded and captured by Union Capt. Nathaniel Lyon. No shots were fired and no resistance offered, but when a local column of Union Home Guards marched the 892 prisoners through St. Louis, an angry crowd gathered. Shooting began. Twenty-eight people were killed and seventy-five wounded. This incident, or "massacre" as pro-southerners called it, set off Missouri's civil war. Duke, *History of Morgan's Cavalry*, pp. 44–50; James W. Covington, "The Camp Jackson Affair: 1861," *Missouri Historical Review* 55 (April 1961): 197–212.

to activate a camp for six of its companies (to meet 21 May). The actions of the KSG had two purposes: first, to guard against Union military actions as had just occurred in St. Louis; and second, to ensure public tranquility as the neutrality proclamation became known.[30]

Working together, Magoffin and Buckner now attempted to restructure the commonwealth into an armed neutral, positioned to repel any invaders from the North or South. Under Magoffin's direction, his state government energetically launched a neutral "foreign" policy, sending emissaries to Lincoln and Davis as well as to Union and Confederate military commanders. Two-man diplomatic teams, consisting of a pro-Union and a pro-southern negotiator who had pledged to promote Kentucky's neutrality and interests, enjoyed initial successes, securing guarantees that Kentucky would not be invaded by either side.[31]

However, by June, the situation in Columbus, Kentucky, a hotbed of secessionist sentiment, threatened to destroy Magoffin's and Buckner's attempts to enforce neutrality. The town boasted a number of Confederate flags and banners, which invited Union gunboats to threaten the town with naval gunfire. Outside observers labeled Columbus "Kentucky's Charleston." To squelch secession sentiment, Buckner, on 10 June, in the most unusual act of his KSG inspector generalship, ordered six companies of the 4th Battalion (KSG troops from Paducah) to deploy to Columbus in order to pacify secessionists and to enforce state neutrality.[32] He placed Lloyd Tilghman in command and ordered Capt. Henry Lyon of the engineer corps to join them. Buckner had been busily engaged throughout western Kentucky at this time. Prior to his activation of the KSG troops, he had persuaded Confederate States of America Brig. Gen. Gideon Pillow to cancel a planned Kentucky invasion, but Columbus' problems and the threat from Union gunboats required decisive action. According to Buckner, "the highly excited state of the citizens of Columbus and vicinity, and the indiscretion of many of them, at every moment imperiling the peace of the Commonwealth, induced me to . . . call into the field a small military force [whose object is to] quiet the unhealthy excitement."[33]

Buckner clearly stated that his purpose in activating the KSG was to "protect all citizens" and to "carry out the obligation of neutrality which the State has assumed . . . restraining our citizens from acts of lawless aggression." One newspaper was shocked, asserting that the troops had been called out "to protect Union men." In ordering this most peculiar of actions, Buckner, a southern rights man, had deployed pro-southern militia units to quell secession

---

[30] *Paris Western Citizen*, 31 May 1861; KSG Special Orders 126, 127.

[31] Searching for additional weapons, Magoffin also sent purchasing agents to the North and South.

[32] Buckner activated 4 infantry companies, 1 artillery company, and 1 cavalry company to move from Paducah to Columbus.

[33] *Louisville Daily Courier*, 12 Jun 1861.

sentiments and uphold Kentucky neutrality![34] Why had he done this, and to what purpose?

Setting Traps

Unionists believed, and Buckner's actions and those of his subordinates seem to suggest, that secessionists desired to keep Kentucky neutral as a first stage or half-step to disunion until a majority of Kentuckians finally made up their minds that their true destinies lay with the South. Conversely, Buckner and other disunionists must have been extremely discouraged as they witnessed the creation of opposition Union Home Guard militias equipped with thousands of Lincoln Guns put into the hands of loyal Union men. How could they dare hope, as many privately whispered, to "take Kentucky out?"

Providentially, the overt and aggressive actions of Union Capt. Nathaniel Lyon at St. Louis on 10 May provided KSG conspirators with a usable template for revolution—an incident similar to St. Louis, if it were to occur somewhere in Kentucky, would allow disunionists to rally an outraged Bluegrass state to the southern side. "If Unionism means such atrocious deeds as I have witnessed in St. Louis, I am no longer a Union man," a Missourian who had strongly opposed secession exclaimed. Many citizens of Kentucky shared his thoughts on the subject. Kentuckians were edgy.[35]

Could Kentucky Unionists be provoked, trapped, or manipulated into perpetrating an act of violence on Kentucky soil? Southern rights military men hoped so. For years, Kentuckians had heard that the "Black" Republicans were aggressive abolitionists who would stop at nothing. Following St. Louis, the belief that federal usurpations formed part of a larger Republican conspiracy to subjugate the border slave states gained dominance. Alfred Pirtle's friend Cabell from St. Louis believed,

> That [Missouri] will be changed by these high-handed actions into secession and then the Federal government having succeeded in their object of precipitating the State will throw so many and such large bodies of troops into the State that the citizens of Missouri will find themselves overawed and held in check by the hands of hireling Abolitionists . . . he sees in the Administrations movements towards our Commonwealth [Kentucky] indications of such proceedings here.

That Pirtle, later a Union Army officer, shared the same outrage and conspiracy beliefs as most Kentuckians, is seen in his comment, "We hope the time will not find us so unprepared as Missouri was."[36]

---

[34] Ibid., 15 Jun, 1 Jul 1861.

[35] Christopher Phillips, *Damned Yankee: The Life of General Nathaniel Lyon* (Columbia: University of Missouri Press, 1990), p. 193; Pirtle, Journal, 11, 15 May 1861.

[36] Pirtle, Journal, 15 May 1861.

Believing Lincoln and Republicans to be aggressive by nature, apparently all Buckner and his KSG cohorts need do was set the traps. And so, beginning in May 1861, this is exactly what Buckner and the KSG attempted to do. They ordered the KSG militia into a number of camps across the commonwealth in hopes that Unionists would attack at least one of them. One Union attack on a KSG encampment, regardless of the military outcome, would act as a catalyst for revolution. "Indeed, the Secessionists of the State Guard, if there be any, went out on purpose to be taken, perhaps," an embedded reporter observed. "They will hold Lincoln to be meaner than ever if he doesn't accommodate them in this cherished wish of their gizzards." Still, "it is rather ominous that a cause needs blood to give it vitality."[37]

With a potentially hostile Union Army camp just across the Ohio River from Louisville (Camp Joe Holt), KSG Lt. Col. Thomas H. Hunt carefully chose an exposed campsite. Expecting a federal attack, Hunt decided to train his battalion at Shepherdsville, thirty miles south of Louisville. Positioning his camp at a bend on the south side of the Salt River, Hunt began training his men in six-day iterations. He named the site Camp Shelby. The Paroquet Springs resort, conveniently at hand, lent the encampment the same social-military atmosphere that had prevailed at the 1860 encampment.

Hunt went into camp with six companies of his regiment on 21 May for one week of training. He expected the Kentucky Rangers (cavalry) and the Citizens Artillery in a few days. To read Citizen Guard soldier Pirtle's account of camp life, one would think that the entire enterprise consisted of sheer boredom. According to Pirtle, "The hours are spent reading, writing, card-playing, rowing on the salt river." In fact, "reading and lounging around is the order of the day."[38]

Magoffin and Buckner perhaps thought otherwise. Convinced that Unionists were about to move, they awaited action. Receiving what later proved to be false reports, Magoffin informed Buckner that he believed that a Union force from Cincinnati would attack Camp Shelby between 21 and 26 May. Events in Missouri filled everyone's minds. Rumors of an imminent attack circulated. "It has been softly whispered," wrote reporter Charley Kirk, "that if this camp is continued (and the probability is it will be for some time) the 'Abolitionist Administration' will adopt the same measures in regard to it that they did so effectually with the St. Louis Brigade. . . . We have an eye to this."[39]

Buckner arrived at Camp Shelby to take command on 26 May and the following scene ensued: "Last evening was one of excitement in our little camp. A rumor was set afloat that dispatches of great importance had been received at Headquarters [and when] orders to sleep on arms and 40 rounds

[37] *Daily Louisville Democrat*, 28 May 1861.

[38] KSG Special Orders 127; Pirtle, Journal, 21, 15 May 1861.

[39] *Louisville Daily Courier*, 27, 30 May 1861.

of cartridges were issued the boys gave vent to their feelings in three cheers. Picket Guards were posted last night."[40]

Forty rounds of ammunition was standard combat issue in 1861, for Hunt's battalion state neutrality had become mighty peculiar. But the awaited Union attack never came—the traps of the KSG had all been set in vain. Union leaders of Kentucky had also observed the events in Missouri and had learned the appropriate lessons. Kentucky Unionists opted to await events while building a Union Home Guard as a deterrent force. Meanwhile they sought bloodless ways to eliminate the KSG.

Dismantling of the Kentucky State Guard *"Some Have Gone to Parts Unknown"*

The end of the KSG came about by cutting off funds, redistributing weapons, and requiring loyalty oaths. Suspicious as ever, the UD-dominated State General Assembly demanded access to Magoffin's correspondence and transactions with the Confederate government and insisted that all KSG militiamen take an oath of loyalty to the United States. It also arranged to divide the weaponry between the KSG and the Home Guard units. Finally, on the last day of the May session, the legislature set up a five-member military board to oversee the arming of Kentucky. "Humiliating as it is," a southern rights man noted, this creation stripped Magoffin of "all his military power."[41]

The results of the Union-dominated military board's decisions did not take long to effect a change. Secessionists in the KSG, tired of marking time while war raged about them, slowly but surely left the KSG ranks and went south to join the Confederate Army—many of them turning over their arms to Home Guard units in their communities. A lack of funding would force the closure by mid-July of Camp Joe Daviess—a permanent training camp established by Hunt in early June atop Muldraugh's Hill—and other places like it.[42]

The Union loyalty oath proved to be most deadly. The insistence of southern rights men upon a code of honor was admirable but naïve for revolutionaries. They again played into Unionist hands. In this instance, Kentucky Unionists understood southern rights men better than southern rights men understood themselves. By insisting on a Union loyalty oath, pro-southern members of KSG units quit their ranks in large numbers.[43]

---

[40] Ibid., 28 May 1861.

[41] Pirtle, Journal, 10 May 1861; E. Merton Coulter, *The Civil War and Readjustment in Kentucky* (Chapel Hill: University of North Carolina Press, 1926), p. 87; *Louisville Daily Courier*, 22 May 1861. Quote from *Frankfort Commonwealth*, 29 May 1861.

[42] At first, $5,000 in training funds had been authorized (15 June), and $30,220 for powder, caps, muskets, balls and shot, lead, and musket repair. By July, the military board would undo this decision.

[43] *Official Records*, ser. 3, vol. 1, p. 238; Ltr, McClellan to Townsend, 17 May 1861.

Lucas G. Hughes informed Governor Magoffin, "The members of the Hancock Rifle Company K.S.G. in the 2nd Saturday in July 1861 at their Company meeting, after reading of the General Order No. 4 refused to take the oath required . . . having thereby become disbanded." His letter was one of many. Submitting his resignation, M. S. Kouns admitted that his company's strength had fallen to seventeen members: "Some have Vol[unteered] in the Federal Army & Some have gone to parts unknown."[44]

To most modern observers, the oath appears trivial, but, to many men of 1861, matters of principle and honor defined themselves. Pirtle understood the issue and was alarmed. He worried that if the oath was insisted upon, "the only arm the State now has would be disbanded."[45]

A few pro-secessionists dodged the oath. They understood the object of Unionists and urged their comrades not to feel obligated by having sworn. Pirtle was not impressed when one of Buckner's aide-de-camps, Maj. Alexander Cassedy, dropped by the Citizen Guards to administer the oath. "A great diversity of opinion exists as to the obligation imposed by the oath, some taking it very lightly," Pirtle noted. "The officer administering it, Cassidy [*sic*] said he would be willing to take it every morning before breakfast thus speaking lightly of the oath."[46]

Money, state armaments, and loyalty oaths—none of this eventually mattered, for on 21 July, as the news of the Battle of Bull Run became known, Buckner and his loyal but pro-southern officers would resign their positions in the Kentucky State Guard and go south to join the Confederate Army. The KSG was at an end. Union Home Guard militias now held the high ground—they had saved Kentucky for the Union.

Conclusions

While the KSG existed, its contribution to Kentucky in 1861 was significant in many respects. Buckner later argued that his pro-neutral KSG delayed a Union and Confederate invasion while preserving the peace of Kentucky. The record shows Magoffin forcefully advocated the use of the KSG as an instrument of neutrality and that he remained, at best, a secret secessionist. Indeed, when examined over time, events seemed to have forced Magoffin to evolve into a sincere neutralist.

Almost as importantly, the KSG recruited Kentucky secessionists and held them in check. It inadvertently paralyzed the revolutionary zeal of men ready to join the rebellion and force Kentucky out of the Union at the point of the bayonet. At camps such as Joe Daviess on Muldraugh's Hill, they dreamed of secession and glory and waited for orders and military action that never came.

---

[44] Ltrs, L. G. Hughes to Magoffin, 21 Sep 1861; M. S. Kouns to Magoffin, 3 Sep 1861, both in Governor's Military Correspondence, Kentucky Historical Society.

[45] Pirtle, Journal, 18 May 1861.

[46] Ibid., 3 Jun 1861.

The KSG saved Louisville from a possible Blanton Duncan secession plot in April, and in June it pacified the would-be secessionists of Columbus. When southern rights men burned a Kentucky Central railroad bridge near Cynthiana to stop the southward movement of Lincoln Guns in August, Magoffin granted Thomas L. Crittenden permission to call out a KSG company if needed. At every turn, the KSG, despite its pro-southern proclivities, had helped preserve the hegemony and peace of Kentucky.

The KSG's second contribution was strategic. Albeit unwillingly at times, it backed state neutrality with force. The presence of a well-armed pro-southern militia willing to back neutrality kept Unionists temporarily off-balance. Tacticians on both sides had to abide by neutrality rather than risk a bloodbath. So long as a sizable portion of the KSG remained in Kentucky and loyal to Magoffin-Buckner, neutrality stood a chance. Buckner fully understood the KSG was the only force in Kentucky that included southern rights, neutrality, and Union men in its ranks. Only with the dismantling of the KSG could the commonwealth take a decisive stand for the Union. Meanwhile, the hollow force stood as a potential nuisance to invaders—perhaps just enough of one to discourage belligerents early in the war.

The third contribution of the KSG was political. KSG companies bolstered and supported the Southern Rights Party at meetings and political gatherings. Without KSG protection, Union men may have broken up southern rights meetings. The odds are that SRP leaders would have been subjected to arrest or forced to flee Kentucky much sooner then September 1861. In this sense, the KSG added to the longevity of Kentucky's Southern Rights Party, giving it backbone and allowing the debate over North or South to continue well beyond that of any other southern state.

Finally, the Kentucky State Guard and Union Home Guard made possible a more peaceful process of polarization between UD and SRP constituencies. Unionists joined Home Guard units, while SRP men joined KSG companies. Kentuckians sorted themselves out peacefully, and over time the extremists of both sides were siphoned off to rival Confederate or Union armies gathering on the commonwealth's borders instead of fighting it out inside the state. The KSG, in ways unique and unforeseen, had helped assist in preserving state hegemony, internal peace, and political freedom. Be it Kentucky 1861 or Iraq 2008, peaceful or violent, militias have their uses.

The Spoliation of Defenseless Farmers and Villagers: The Long-Term Effects of John Hunt Morgan's Raid on an Indiana Community

Stephen I. Rockenbach

On the morning of 9 July 1863, a group of horsemen surrounded a two-story farmhouse situated along the Mauckport road leading from the Ohio River to the town of Corydon, Indiana. One of the men dismounted and entered the house, setting the furniture on fire. Catherine Glenn tried to dissuade the intruder, while her husband, Peter Glenn, came running down the stairs, followed by his adult son, John. Peter wrestled with the invader while John attempted to extinguish the fire. A second raider entered the house and ended the scuffle by shooting Peter Glenn, who stumbled outside into the yard. John Glenn rushed to his father's side, but the horsemen shot the young man through both thighs. The raiders left the mortally wounded Peter Glenn, his seriously injured son, and their horrified wives to watch the house burn and headed toward the town of Corydon. Although this incident was part of Confederate General John Hunt Morgan's infamous "Great Raid" through Ohio and Indiana, it shows evidence of the same brutal tactics used by Civil War guerrillas throughout the upper South. Why would troopers of the 5th Kentucky Cavalry—not guerrillas in the true sense of the word, but regular troops—commit such a violent action against civilians? Their motivation becomes clearer once we realize that, early that morning near the Glenn house, cavalrymen from the 6th Regiment of the Indiana Legion (the state militia organization) charged a scouting party of forty Confederate cavalry and killed Pvt. John Dunn of the 5th Kentucky. His fellow Kentuckians burned the Glenn farm as an act of vengeance perpetrated by men who entered the state of Indiana intent on bringing the war north of the Ohio River. Indeed, they caused a considerable amount of economic loss to civilians and inaugurated two years of guerrilla violence and partisan warfare in a region previously spared such depredations.

Many historians, amateur and professional alike, enjoy retelling Morgan's exploits, although the celebratory nature of these accounts obscures the overall effect of irregular warfare on civilian populations.[1] My approach departs

[1] The literature pertaining to Morgan's July 1863 raid is vast and ranges from voluminous collections of stories to short, county-specific descriptions of events. See the following books for various, and often conflicting, accounts of the raid, including some of the incidents that occurred in Corydon and Harrison County: Lester V. Horwitz, *The Longest Raid of the Civil*

from the raid narrative and incorporates the approach of social history and community studies. Studying the raid's effect on one community reveals how the different forms of irregular warfare—partisan warfare, guerrilla tactics, and cavalry raids—intertwined along the Ohio River border. The material effects of the raid were significant, perhaps more than we could comprehend with mere numbers, but, after the raid, citizens in southern Indiana also became disillusioned with both their Kentucky neighbors (many of Morgan's men, and Morgan himself, were Kentuckians) and the Union military authorities (whom they blamed for allowing Corydon to be occupied in the first place). On that hot July day in 1863, as Peter Glenn lay bleeding in front of his burning house, Corydon began the transformation from river valley community to besieged border town.

Southern Indiana's first indoctrination into border war was significant because it occurred as the war steadily hardened. During the first two years of the war, most military commanders avoided targeting civilian populations in favor of achieving military objectives, but guerrilla fighters consistently used retribution, coercion, or confiscation to achieve their means. In reference to "hard war" strategy, historian Mark Grimsley states that in early 1863 Union commanders had not "yet harnessed those energies in major operations against Southern infrastructure and society." Morgan, not a guerrilla leader in the true sense, was one of a number of Confederate commanders who Grimsley concludes "fought in unorthodox but mostly above-board fashion."[2] Additionally, Robert R. Mackey demonstrates the complexity of this topic by providing useful categories for the three different types of irregular warfare common to the upper South during the Civil War: "guerrilla, or people's war, partisan warfare, and conventional cavalry used as raiders." Although Mackey's study highlights particular aspects of irregular warfare, he notes that "all three types of irregular warfare existed simultaneously throughout the Upper South, with a varied amount of impact on the Federal forces." Mackey concludes that irregular warfare proved fruitless for the Confederacy, because the Union Army successfully developed countermeasures.[3] However, if viewed from the perspective of border citizens, cavalry raids and guerrilla warfare did have significant and enduring effects. In the case of Corydon, the Morgan

War: Little-Known and Untold Stories of Morgan's Raid into Kentucky, Indiana, and Ohio, rev. ed. (Cincinnati, Ohio: Farmcourt Publishing, 2001), passim; David L. Taylor, *With Bowie Knives and Pistols: Morgan's Raid in Indiana* (Lexington, Ind.: Taylormade Write, 1993), passim; Arville Funk, *The Morgan Raid in Indiana and Ohio* (Corydon, Ind.: ALFCO Publishing, 1971), passim.

[2] Mark Grimsley, *The Hard Hand of War: Union Military Policy Toward Southern Civilians, 1861–1865* (New York: Cambridge University Press, 1995), p. 105 (first quote) and p. 112 (second quote).

[3] Robert Russell Mackey, *The Uncivil War: Irregular Warfare in the Upper South, 1861–1865* (Norman: University of Oklahoma Press, 2004), p. 5 (first quote), p. 6 (second quote), and see also pp. 7–9.

raid shattered people's sense of security, encouraged further violence, and crippled the local economy.

When Morgan's raiding party of approximately 2,500 cavalry and 4 artillery pieces occupied the hills above Brandenburg, Kentucky, on 7 July 1863, Union authorities knew of the Confederate military presence in Kentucky. However, Army commanders failed to prevent Morgan's force from reaching the Ohio River. On 8 July, Union officers throughout the river border learned that Morgan had taken Brandenburg and had begun crossing the Ohio River using two captured steamboats.[4] A handful of local forces and one Union gunboat failed to halt Morgan's crossing. Some of the residents of the southern portion of Harrison County fled to Corydon with whatever property they could carry. As citizens armed themselves, hid valuables, and prepared for the worst, the rebel cavalry took food, tack, livestock, fodder, and whatever else they wanted from the houses and villages in the area.

Southern Indiana's official military branch, the Indiana Legion, responded to the threat by mobilizing its members and any willing volunteers. However, the legion in Harrison County mustered approximately five hundred men, while well over two thousand Confederate cavalry approached the town. On 9 July, the legion cavalry spread out along the roads south of town and prepared to ambush the Confederate attackers, while the legion infantry companies constructed a barrier on the southern edge of town. The townspeople knew that they could not stop the larger enemy force, but they hoped to slow down the approaching enemy and give the legion troops from the nearby city of New Albany time to arrive.[5] Writers and some local enthusiasts refer to the thirty-minute fight that preceded Morgan's capture of the town as the "Battle of Corydon," even though participants and military authorities always referred to the incident as a "fight" or a "raid." The Confederate troops made one poorly conceived charge against the entrenched legion companies before encircling the defenders. The legion and its volunteers retreated into town, but they surrendered after Morgan's artillery fired on several buildings occupied by women and children.

Perhaps something more akin to a full-fledged battle would have occurred had reinforcements from the nearby city of New Albany arrived. Brig. Gen. Jeremiah T. Boyle, who commanded the state and federal troops in the Louisville area, believed that Morgan intended to attack New Albany and, as a result, refused to let the city's legion companies go to the aid of their

---

[4] Brev. Lt. Col. Robert N. Scott and Maj. George B. Davis, *The War of the Rebellion: A Compilation of the Official Records of the Union and Confederate Armies*, 128 vols. (Washington, D.C.: Government Printing Office, 1880–1898); Brig Gen Asbroth to Maj. Gen. Burnside, 8 Jul 1863, ser. 1 sol. 23, pt. 1, pp. 709–10.

[5] Rpt, William Farquar, in *Operations of the Indiana Legion and Minute Men, 1863–4* (Indianapolis: W. R. Holloway, State Printer, 1865), pp. 42–43; Thomas Slaughter to "Sir," 17 Aug 1863, Indiana Legion Records, folder 5, box 10, Indiana State Archives (ISA), Indianapolis, Ind.

comrades in nearby Corydon. After the raid, William Hisey, the Harrison County treasurer, wrote to Indiana Governor Oliver P. Morton asking for General Boyle's resignation on the grounds that the Union commander let "rebel cavalry, with artillery, destroy a country town."[6] Although Hisey spoke with the bitterness of a man who lost $786.87 in cash and property to the raiders, the complaint was not totally unfounded.[7] Boyle's inactivity in part explains why the Indiana Legion proved so ineffective during the crisis. For Corydon's citizens, the raid exposed the flaws in Indiana's militia organization and started a gradual decline in faith in Union military authority.

The immediate and extensive effects of the raid, including eight Harrison Countians killed and several wounded, threatened the community's livelihood. The raiders stole or damaged over $80,000 in property in Harrison County alone. Horses and tack were the most common items taken, although Morgan's guerrillas helped themselves to valuables, food, cash, and clothing. Merchants suffered the highest monetary loss, including Samuel J. Wright, who lost approximately $5,524 in merchandise from his store in Corydon.[8] His losses forced Wright to place an ad in a local paper asking all his customers to pay their debts to him because "Morgan's band of thieves [had] robbed me of at least half my goods."[9] At the beginning of the war, Wright accepted the potentially lucrative position of quartermaster for the 6th Legion regiment. In January 1864, he resigned as quartermaster, citing the financial burden of the post, supply difficulties, and the raid's effect on his business. Wright eventually sold his store and fell back on his legal training.[10]

Farmers, laborers, mechanics, and tradesmen who lost far less monetarily than Wright also risked privation, especially those farmers trying to finish the

---

[6] H. S. Hisey to Gov Morton, 23 Jul 1863, microfilm reel 6, O. P. Morton Papers, ISA.

[7] James Ramage considers the $690 in cash that raiders took from William Hisey to be "Union funds." James A. Ramage, *Rebel Raider: The Life of General John Hunt Morgan* (Lexington: University Press of Kentucky, 1986), p. 172. Regardless of whether or not Morgan justified this behavior as confiscating property from the Union government, Hisey requested compensation for the stolen money and other valuables. The United States Quartermaster Department (USQD) denied the claim in 1886, and Hisey was never reimbursed. File 248, box 603, Record Group (RG) 92, Misc. Claims, 214, National Archives and Records Administration (NARA), Washington, D.C.

[8] Indiana Morgan Raid Claim Commission Records, microfilm, p. 37, Indiana Historical Society Library (IHS), Indianapolis, Ind.

[9] *Corydon Weekly Union*, 10 Nov 1863, reprinted in News of Long Ago: From Republicans of Another Century scrapbook, Corydon Local History and Genealogy Library, Corydon, Ind.

[10] S. J. Wright to A. Stone, 18 Jan 1864, folder 5, box 10, Indiana Legion Records, ISA. The raiders purposely targeted other businessmen, charging a $500 protection fee (under threat of burning down the buildings) for three of the community's mills. In the 1880s, the owners of the three mills applied for reimbursement for the ransom they paid Morgan. Additionally, Phillip Lopp asked for $4,761 for his mill, which raiders suspected as housing bushwhackers and subsequently burned on 8 July 1863. None of these claims fit the specifications of the Quartermasters Act of 4 July 1864; therefore, the USQD refused payment. File 119, box 602; files 134 and 203, box 603; and file 462, box 605. All in RG 92, Misc. Claims, 214, NARA.

wheat harvest. Many citizens lost horses, essential for many of the jobs related to harvesting and marketing surplus goods.[11] The raiders took between 300 and 400 horses in Harrison County, while the Union cavalry in pursuit of Morgan took an additional 150 mounts, often without leaving proper receipts with the civilian owners.[12] A number of male citizens stayed to hide their property instead of joining the legion companies in town, while others left the job to their wives and children. Isaac Pitman returned to his fields the day after the raid to cut wheat with his three horses and one borrowed animal. A group of raiders had overlooked the animals when Mary Pitman made them breakfast on 9 July, but the following day Union cavalrymen took two horses and left Pitman to finish his work with half a team.[13]

State and military authorities took steps to remedy the raid's effect on local agriculture, although not all of the regulations regarding recovered horses worked in favor of the farmers. Governor Morton ordered General Henry B. Carrington, commander of the Indiana military district, to issue a detailed description of the procedure for reporting and recovering stolen or confiscated horses. The message detailed the policies and procedures needed to remedy "the exigencies of the harvest and the interruption of the farming interests by the John Morgan raid." Carrington ordered all citizens who had replacement horses to give these animals to the provost marshal. Farmers who lost horses were able temporarily to keep such found horses, but only until the end of the harvest. The provost marshal gathered all the recovered horses in each locality and redistributed them to farms, depending on individual need, a measure that guaranteed that citizens could bring in the wheat harvest and transport it to market. However, once done, citizens who did not turn in animals were open to prosecution.[14]

Civilians encountered a number of problems with the procedure for claiming animals and applying for reimbursement. Those citizens who lost horses to Morgan's men filed descriptions and affidavits, often having to travel as far as Cincinnati (approximately a hundred miles away) to find their animals. The United States Quartermaster Department (USQD) did not attempt to return animals that Union forces had captured, offering them for sale instead. Even people who lost horses to Union cavalry did not always have an easy time getting reimbursement. The USQD required the claimant to produce a receipt or at least two sworn witness statements indicating that Union forces had taken the property. In the confusion and urgency of the pursuit of Morgan,

---

[11] J. Sanford Rikoon, *Threshing in the Midwest, 1820–1940: A Study of Traditional Culture and Technological Change* (Bloomington: Indiana University Press, 1988), pp. 14–16.

[12] Henry Beebee Carrington to Gov O. P. Morton, Filson Historical Society (FHS), Louisville, Ky.; Files 123, 147, 171, box 603; files 265, 342, 345, box 604; files 405, 424, 459, box 605; and file 503, box 606. All in RG 92, Misc. Claims, 214, NARA.

[13] Files 453 and 434, box 605, RG 92, Misc. Claims, 214, NARA.

[14] "Losses and Impressment of Property during the Morgan Raid: General Carrington Order," 16 Jul 1863, O. P. Morton Papers, microfilm reel 6, ISA. See also *New Albany Daily Ledger*, 17 Jul 1863. For the myth about Indiana and Ohio farmers benefiting from the superior horses left by Morgan's raiders, see Horwitz, *The Longest Raid*, p. 377.

many claimants did not get valid receipts, and those who did could not always locate them. Henry Richard was standing in his farmyard watching Union cavalrymen when one of the troopers approached, carrying his saddle and bridle. The soldier took Richard's mule to replace the cavalryman's recently expired mount but gave no receipt in turn.[15] Although civilians were elated to see the Union troopers, the pursuit further burdened the community.

At its essence, the raid into the border free states was political in motivation and design. Morgan wanted to bolster the hopes of Kentucky secessionists while making Hoosiers and Buckeyes, quite literally, pay for their states' role in supporting the Union's military and political efforts to keep secessionists from gaining control of Kentucky. The *New Albany Daily Ledger* declared that Morgan violated "all the rules of civilized warfare" by robbing citizens, extorting civilian businesses, and stealing horses. The newspaper complained that Morgan did little damage to railroads or other military objectives, instead focusing on the "spoliation of defenseless farmers and villagers."[16] Affected civilians could not be convinced that Morgan's raid had any strategic objective. Instead, the raid was a direct attack on themselves, their property, and their livelihood.

The July 1863 raid soon encouraged small guerrilla bands to pick up where Morgan left off, launching raids along the river border and forcing border residents to dedicate their energy to protecting their homes and property instead of aiding Kentucky Unionists. Yet the guerrilla presence did not necessarily draw troops away from the front or help the Confederate war effort in any significant way. The raid inaugurated a period of guerrilla activity along the Ohio River and emboldened the activities of local partisans, who used the opportunity to exact vengeance upon their Unionist neighbors. These small independent bands of raiders, regardless of whether they claimed legitimacy under the Confederacy's Partisan Ranger Act, applied Morgan's methods on a local basis.[17] Citizens in southern Indiana feared that Morgan's success encouraged "men made desperate by the dark clouds now overhanging their sinking cause" to assault and rob Unionists living north of the Ohio River. Community leaders urged civilians to defend themselves and bitterly reminded border residents that military authorities had failed to help them in early July when "the wolf was upon us."[18] Morgan's raid,

---

[15] Files 435 and 455, box 605, RG 92, Misc. Claims, 214, NARA.

[16] *New Albany Daily Ledger*, 23 Jul 1863.

[17] Scott J. Lucas, "'Indignities, Wrongs, and Outrages': Military and Guerrilla Incursions on Kentucky's Civil War Home Front," *Filson Club History Quarterly* 73 (October 1999): 371; James B. Martin, "Black Flag over the Bluegrass: Guerrilla Warfare in Kentucky, 1863–1865," *Register of the Kentucky Historical Society* 86 (Autumn 1988): 352–75. For a discussion of the variety of guerrilla activity occurring in Kentucky during the war, see B. Franklin Cooling, "A People's War: Partisan Conflict in Tennessee and Kentucky," in *Guerrillas, Unionists, and Violence on the Confederate Home Front*, ed. Daniel E. Sutherland (Fayetteville: University of Arkansas Press, 1999), pp. 113–32.

[18] *New Albany Daily Ledger*, 20 Jul 1863.

combined with subsequent guerrilla activity along the river border, brought the war home to Harrison Countians.

The fluidity of the border war explains how southern Indiana, a relatively peaceful region until the spring of 1863, quickly devolved into a borderland of violence and destruction. The evidence demonstrates a clear connection between Morgan's command and the small bands of guerrillas that terrorized Unionists: most of the "war rebel" or "home rebel" leaders had once belonged to Morgan's cavalry. B. Franklin Cooling's research supports the idea that guerrilla war was fluid and difficult to define, noting that in Tennessee and Kentucky the separate levels of people's war "often blurred in defiance of easy interpretation." The people's war included guerrilla war, but could also encompass fighting between militia and volunteer armies.[19] This was the situation in the vicinity of Corydon, where the legion struggled to fend off pro-Confederate forces, and prominent Unionists became targets for violent retribution.

In 1864, guerrillas in Meade County and other Kentucky counties near the Ohio River intensified their attacks against Unionists in the months leading up to the local elections that fall. The guerrilla bands wanted to intimidate, kill, or run off local Unionists in order to establish local political control.[20] Brandenburg, the seat of Meade County, was the epicenter of guerrilla activity. Between June 1864 and February 1865, partisans perpetrated several raids, robberies, and shootings in the vicinity. Many of these men were from Meade County, and they intimidated, murdered, and robbed their Unionist neighbors.[21] In July 1864, seventeen guerrillas crossed the river on skiffs, entering the southeast corner of Harrison County. Five of these men took cash and property from a well-known Unionist, but his neighbors learned of the disturbance and helped rout the bandits. The following month, the gang of guerrillas contented themselves with remaining on the Kentucky shore and firing on steamboats passing through the area. These attacks were not random violence but rather the result of the guerrillas' "bitter hate" for meddling Unionists living north of the river.[22] These individuals had given up on affecting the outcome of the war and resolved to discourage Unionists in southern Indiana from influencing politics in Kentucky. In effect, they followed Morgan's example by heaping retribution on these primarily Unionist communities.

One solution to the guerrilla problem was to reaffirm the alliance between Unionists in southern Indiana and Kentucky. Brig. Gen. Henry Jordan, commander of the Indiana Legion, wrote Indiana's Adjutant General, William H. H. Terrell, in November 1864, outlining the issues facing Corydon's Unionist population and proposing a plan to remedy the violent conditions on

---

[19] B. Franklin Cooling, "A People's War: Partisan Conflict in Tennessee and Kentucky," p. 113.

[20] *New Albany Daily Ledger,* 16, 21 Jun, 9 Aug 1864.

[21] Lizzie Schreiber Diary, Stith-Moreman Papers, FHS; *New Albany Daily Ledger*, 30 Jun, 12, 17, 19 Aug 1864, and 4 Jan, 14 Feb 1865.

[22] *New Albany Daily Ledger*, 27 Jul, 9 Aug (quote), 19 Aug 1864.

the border. He described how the guerrillas often fired at Harrison Countians from the Kentucky shore or discouraged citizens from going to Kentucky to conduct business or visit friends and family. Jordan suggested neutralizing the threat by raising one or two regiments of cavalry for six months' service on the south side of the river. Men from southern Indiana would join these units, which, according to Jordan, would compel the soldiers to take their duty seriously and to refrain "from wanton depredations on property."[23] Governor Morton, a Republican, never supported this plan and spent more time suppressing Democratic political opposition than tending to the problems facing the state's southern border.

Morgan's raid on Corydon had an enduring economic effect, although the true extent of the damage only becomes apparent when one looks into the postwar army records and newspaper articles. In 1867, some 468 Harrison County residents filed $86,551.72 in claims with the U.S. government, a sum of which totaled about 5 percent of all personal property listed in the 1860 census.[24] These citizens required quick compensation in order to replace the animals and tack essential to rural life, but the official response to their losses was slow and indecisive. Indiana's state government was responsible for passing legislation for payment of these claims, with the understanding that the federal government would eventually reimburse it. But Indiana's bitter political partisanship prevented victims of Morgan's raid from receiving the swift action they needed to replace property and pay debts.[25]

For many Harrison County citizens, the ordeal shaped their postwar experience and diminished their faith in the state and national governments that failed to protect them from danger or compensate them for their losses. During the 1880s, the United States quartermaster general (USQG) tried to review Indiana's outstanding claims, but the Indiana state government failed

[23] Rpt, Brig Gen Henry Jordan, in *Operations of the Indiana Legion and Minute Men*, pp. 102–04.

[24] Report of Morgan Raid Commissioner to the Governor, 31 Dec 1869, RG 92, Collective Correspondence file, Morgan's Raid, box 697, NARA; *U.S. Historical Census Data Browser*, accessed 28 Nov 2004.

[25] U.S. Quartermaster General's Office to H. S. Morey, 4 Apr 1882, RG 92, Collective Correspondence file, Morgan's Raid, box 696, NARA. Lester Horwitz incorrectly concludes that Indiana paid its claims in the same fashion as Ohio, which assessed and settled all claims at the end of the war (Horwitz, *The Longest Raid*, pp. 60, 383–84). In Indiana, Republicans seriously undermined efforts to address the Morgan's raid losses in an attempt to maintain control over the state government. After the elections in 1862, Republican state senators reacted to the Democratic gains in the state assembly by leaving the capital and refusing to allow a quorum. The Democratic senators wanted to take away Morton's ability to control the Indiana Legion, which led Republicans to claim that "Copperheads" were planning to use the state troops to form a northwest Confederacy. Morton not only approved of this tactic, but he operated the state with private and national funds from 1863 through 1865, occasionally using unlawful methods to accomplish this (Emma Lou Thornbrough, *Indiana in the Civil War Era, 1850–1880* [Indianapolis: Indiana Historical Bureau, 1965], pp. 185–87).

to release sufficient documentation. Indiana's adjutant general claimed that the detailed affidavits and documentation that the commission had collected were "lost."[26] Between 1880 and 1887, the USQG assessed and adjudged the claims as best it could, based on resubmitted information from the citizens of southern Indiana, including 468 claims from Harrison County. But the efforts of southern Indiana politicians failed to galvanize support for legislation paying all Morgan's raid claims. Therefore, the USQG processed and paid only those claims for property lost to Union troops, marking all others "taking and use not proven" or "rebel."[27] Harrison County residents and their descendants submitted a total of $81,558.07 of unpaid claims to the USQG, but the amount paid under the Act of 4 July 1864 amounted to only $8,659.[28]

The brief raid on Corydon and Harrison County served as a catalyst for building partisan tensions in the region, but historians assess Morgan's raid as largely ineffective because it did not yield strategically viable results or affect the outcome of military operations. The disruption to Union forces and supply lines that the raid inflicted failed to justify the loss of men and equipment to the Confederate cause. However, the ulterior motives of Morgan and his men included an objective completely detached from Confederate strategy: retribution. Border Confederates wreaked vengeance on the Unionists in the Ohio Valley who kept Kentucky from seceding. If one measures the effect of Morgan's politically motivated "payback" throughout southern Indiana and Ohio, then the raid did have some success. In Corydon, citizens were hard-pressed to affect political events in Kentucky after the raid, although they strongly supported Bluegrass Unionists earlier in the conflict.[29] The raid emboldened pro-Confederate Kentuckians and allowed them to gain local political control during and after the war. As Confederate strategy, Morgan's cavalry raid was impressive, yet ineffective. As an assertion of pro-Confederate

---

[26] Quartermaster General's Office, A Report on the Letter of M. J. Kinney of the 17th, 20 Aug 1880, RG 92, Collective Correspondence file, Morgan's Raid, box 697, NARA.

[27] Samuel B. Holabird, Quartermaster General, to Honorable Secretary of War, 11 Feb 1884, RG 92, Collective Correspondence file, Morgan's Raid, box 696, NARA; *Corydon Democrat*, 29 Aug 1923. There were some attempts to remedy the situation, but they suffered from the same political division that doomed the Indiana Morgan's Raid Claims Commission. In 1884, the Senate Commission of Claims drafted Senate Bill 527 in an attempt to pay all Morgan's raid claims in Indiana and Ohio. Strother M. Stockslager, a lieutenant in the Legion during the raid and a state representative in 1884, worked to gain support for the bill. He believed the bill did not pass because politicians from other states did not have any vested interest in compensating Hoosiers.

[28] These totals were compiled from RG 92, Office of the Quartermaster General, Claims Branch 1861–1889, Quartermaster stores (Act of 4 July 1864), Misc. Claims, bk. 214, claims 36–504, NARA. The total amount of claims submitted to the United States quartermaster general (USQG) differs from the Indiana Morgan's Raid Claims Commission ledger by only $152.83.

[29] Stephen I. Rockenbach, "War upon Our Border: War and Society in Two Ohio Valley Communities, 1861–1865" (Ph.D. diss., University of Cincinnati, 2005), pp. 41–43.

support and determination in the Ohio River Valley, it was significant; the raid expedited the creation of the Ohio River as a barrier between North and South and allowed pro-Confederate forces in Kentucky to assert their influence locally, if not nationally.

Irregular Engineers: The Use of Indigenous Labor in the Rebuilding of Critical Infrastructure During the Korean War, 1950–1953

Eric A. Sibul

Introduction

The Eighth United States Army, Korea (EUSA), made extensive and effective use of indigenous labor to reconstruct the Korean National Railroad (KNR) in autumn 1950 and spring 1951. Indigenous labor carried out the temporary repair of bridges and restoration of railroad lines with remarkable speed, using only native tools and light equipment. Native engineers made bridge repairs using ingenuity and improvised materials. EUSA engineer and transportation officers considered the native work force to be of invaluable assistance in rebuilding railroad lines that were of critical importance for logistic support to combat forces. Although this paper focuses on reconstruction of railroad lines, native labor was also of vital importance to construction and repairs of other critical infrastructure, such as airfields and over 2,700 miles of military roads.[1] Native labor also assisted in ways other than construction: nine companies of A-frame carriers of the Civilian Transportation Corps (later integrated into the Korean Service Corps) carried on their backs food, water, ammunition, and signal and engineering material to EUSA combat troops on mountaintop positions accessible only on foot.[2] This paper will demonstrate how native labor, local organizations, and indigenous skills helped provide logistic support to a modern army.

The Korean National Railroad as a Critical Infrastructure

According to military historian Martin Blumenson, infrastructure had emerged by the time of the Korean War as a decisive strategic concept in

---

[1] James A. Huston, *Guns and Butter, Powder and Rice, US Army Logistics in the Korean War*. (Selinsgrove Pa.: Susquehanna University, 1989), p. 290; Benjamin King, Richard C. Biggs, and Eric R. Criner, *Spearhead of Logistics: A History of the United States Army Transportation Corps* (Fort Eustis, Va.: U.S. Army Transportation Center, 1994), p. 309; Charles W. Voss and Sedgwick R. Bryon, "Aviation Engineers Do the Groundwork in Korea," *Army Information Digest* (February 1953): 28.

[2] Bradley J. Haldi, "Korean Service Corps—Past and Present," *Army Logistician* (July–August 1987): 22–23; Margaret A. Mallman, "Korean Brawn Backs the Attack," *Army Information Digest* (December 1951): 47–49.

military operations. Infrastructure in a theater of operations consists of fixed permanent installations and facilities that make sustained ground force operations immediately feasible. If these installations and facilities do not exist, they must be established before protracted warfare can occur.[3] The KNR was the infrastructure sine qua non for military operations in Korea, 1950–1953. The Japanese built the Korean railroad system between 1904 and 1945 as a strategic network connecting Korean ports to Manchuria. The Korean railroad system was a very well-constructed military railroad almost entirely for the benefit of Japanese forces.[4] The Japanese developed the Korean highway system to serve purely local needs and to serve as a feeder system to the railroads.[5] Given the nature of the Korean transportation infrastructure from 1950 to 1953, the mobility of men, munitions, and supplies in the Korean theater of operations depended on railroads as much as it had in the American Civil War nearly ninety years earlier.[6] Approximately 95 percent of all supplies cleared at ports and moved to forward supply points via the KNR. The same held true in the movement of personnel, including the rotation of troops and the evacuation of casualties.[7] Despite overland movements predominantly less than three hundred miles, motor transport could not supplant railroads in Korea due to the limited road network and the short supply of vehicles, petroleum, and manpower. The number of trucks initially requested for EUSA use in Korea in autumn 1950 was greater than could be obtained either from army depots or from new production. While the KNR had an extensive existing organization, no existing motor transportation organization in the Republic of Korea (ROK) could provide the needed logistic services. Given the come-as-you-are nature of the war, training and organizing a motor transportation service using indigenous labor to supplant the KNR would be too time-consuming and difficult. EUSA did organize three truck companies composed of indigenous civilian labor.

---

[3] Martin Blumenson, "The Emergence of Infrastructure as a Decisive Strategic Concept," *Parameters* (Winter 1999–2000): 39.

[4] Conversation Between General Edmund C. R. Lasher and Lt Col D. R. Lasher, p. 63, Senior Officers Debriefing Program, U.S. Army Military History Institute (USAMHI), Carlisle Barracks, Pa.

[5] Roy E. Appleman, *South to the Naktong, North to the Yalu (June–November 1950)*, United States Army in the Korean War (Washington, D.C.: U.S. Army Center of Military History, 1961), p. 117.

[6] Ollie Atkins and Sylvia Crane Myers, "The World's Worst Railroad Headache," *Saturday Evening Post*, 14 Jul 51, p. 126; William T. Faricy, "Railroads—Mighty Weapon in Korea," *Defense Transportation Journal* (March–April 1952): 30; A. A. Hoeling, "The Army's in the Railroad Business," *American Mercury* (February 1954): 47; Sidney A. Levy, "Build 'em up—Blow 'em up," *Railway Progress* (February 1952): 7; George L. Wiley, "Transportation Corps—A Decade of Service," *Army Information Digest* (August 1952): 32.

[7] Atkins and Myers, "The World's Worst Railroad Headache," p. 126; Hoeling, "The Army's in the Railroad Business," p. 47; Ruben Levine, "Iron Horse vs. Iron Curtain," *Machinist Monthly Journal* (November 1951): 349; James A. Van Fleet, *Rail Transport and the Winning of Wars* (Washington, D.C.: Association of American Railroads. 1956), p. 23.

Based in Pusan, they operated all types of motor vehicles, ranging from jeeps to heavy trucks, gasoline tankers, and wreckers.[8]

KNR was tasked as the primary transportation service in the ROK, but the immense traffic demands quickly overwhelmed its management abilities and the movement control capabilities of the Republic of Korea Army (ROKA), which were both relatively new and inexperienced organizations. Before 1945, railways in Korea were under tight control of the Japanese state. In almost all cases, the Japanese held administrative and technical jobs, while the Koreans for the most part were employed in positions involving mainly manual labor; few held any kind of management position. The positions largely attainable by Koreans were as track and car repairmen, roundhouse hostlers, and passenger car cleaners.[9] From 14 September 1945 to 7 September 1948, the United States Military Government in Korea (USMGIK) operated the Korean railways with the intention of leaving the KNR as a viable, state-owned national railroad for the ROK. Despite largely perfunctory training of Koreans by Transportation Corps personnel to assume technical and managerial positions, the USMGIK seems to have been successful in this effort. Between 7 September 1948 and the outbreak of the war on 25 June 1950, KNR operated with improving efficiency while carrying an increasing amount of freight and passengers.[10] However, the rapid Communist capture of Seoul, where the KNR had numerous facilities, a substantial amount of equipment, and a sizable portion of its work force, as well as the tremendous burden of heavy military traffic and masses of refugees made it impossible for the young railway organization to cope without outside assistance.[11]

Consequently, in July 1950, EUSA negotiated through the U.S. Ambassador to Korea, John J. Muccio, the transfer of operational control of KNR from the ROK government to the United States Army.[12] On 26 August 1950, the U.S. Army activated the 3d Transportation Military Railway Service

---

[8] Huston, *Guns and Butter, Powder and Rice*, p. 180; Julian Thompson. *Lifeblood of War: Logistics in Armed Conflict* (London: Bassey's, 1989), pp. 125–26; Van Fleet, *Rail Transport and the Winning of War*, pp. 26–27; John G. Westover, *Combat Support in Korea*, U.S. Army in Action Series (Washington, D.C.: U.S. Army Center for Military History, 1987), pp. 43–44.

[9] United States Military Government in Korea—Bureau of Transportation, Record Group (RG) 554, National Archives and Records Administration (NARA); G. Harry Huppert, "Korean Occupational Problems," *Military Review* (December 1949): 15; Craford F. Sams, *Medic: The Mission of an American Military Doctor in Occupied Japan and War-Torn Korea* (Armonk, N.Y.: M. E. Sharpe, 1998), p. 206.

[10] *Korea: Its Land, People and Culture of All Ages* (Seoul: Hakwon-sa,1960), p. 266; Kyotongbu [Ministry of Transportation], *Transportation of Korea 1957* (Seoul: Ministry of Transportation, 1957), p. 38.

[11] Neville Brown, *Strategic Mobility* (New York: Praeger, 1964), p. 218; Crump Garvin, "Pitfalls in Logistic Planning," *Military Review* (April 1962): 7.

[12] "Headquarters 3D Transportation Military Railway Service Background," in U.S. Army Forces Far East and Eighth Army, *Logistics in the Korean Operations* (Camp Zama, Japan: U.S. Army Forces Far East and Eighth Army,1955), vol. 3, ch. 6, p. 6, U.S. Army Transportation Museum Library (USATML); Kyotongbu, *Transportation of Korea 1957,* p. 40.

(3d TMRS), which would eventually grow to an organization of two railway operating battalions, one railway shop battalion, and a military police battalion. It also supervised 32,000 KNR civilian employees.[13] Korean personnel were vital, since the 3d TMRS' very much understrength units had no hope of running the railroad system without them. The 3d TMRS worked closely with the Army Corps of Engineers, notably the 32d Engineer Construction Group, in the reconstruction and refurbishment of railroad lines.[14]

Reconstruction Tasks

Indigenous labor was of critical importance in assisting in the reconstruction of the railroad lines after the breakout from the Pusan Perimeter in September 1950; after the Chinese intervention and United Nations Command (UNC) fallback to the Pyongtaek-Ansong-Wonju-Samchok line; and after the advance to the 38th Parallel again in March 1951. With the UNC return to the 38th Parallel in 1951, it regained approximately seventy miles on the central mainline railroad from the defense line to Yongdungpo on the Han River, and twenty-two severely damaged or destroyed bridges. The situation paralleled the situation after the breakout of the Pusan Perimeter in September 1950. With little exception, every bridge rebuilt on the United Nations (UN) advance had been first destroyed by friendly retreating forces.[15] For EUSA, rebuilding bridges offered the greatest single challenge in getting railroad lines back into service. The urgency and immensity of the task required a division of labor. The U.S. Army construction engineers took on the reconstruction of the largest bridges, and the 3d TMRS, mobilizing the resources of the KNR as well as Korean contractors and additional civilian labor, assumed the bulk of the responsibility for repairing smaller railroad bridges, as well as clearing tunnels and re-laying railroad track.[16] Originally, the EUSA paid the KNR monthly for transportation services, and the KNR used the funds to cover repair costs. However, the vast scale of reconstruction exhausted the KNR's funds by December 1950. As a result, the EUSA assumed direct responsibility to pay all labor and other charges associated with reconstruction. Indigenous labor was paid in Korean currency as well as being provided with supplies of food. In the war-torn ROK, the Americans had no problem recruiting

[13] 3d TMRS, Unit History and Activity Report (August–September 1950), RG 407, NARA; *Logistics in the Korean Operations*, vol. 3, ch. 6, p. 6.

[14] Huston, *Guns and Butter, Powder and Rice*, p. 290; M. Clare Miller, "High Steel in Korea," *Military Engineer* (September–October 1951): 332; 3d TMRS, Unit History and Activity Report (1–31 March 1951), RG 407, NARA.

[15] *Logistics in the Korean Operations*, vol. 3, ch. 6, p. 12; Paschal N. Strong, "Engineers in Korea—Operation Shoestring," *Military Engineer* (January–February 1951): 14.

[16] Appleman, *South to The Naktong, North to the Yalu*, p. 639; Huston, *Guns and Butter, Powder and Rice*, p. 290; Joseph R. Slevin, "World's Biggest Traffic Department," *Railway Progress* (October 1952): 9; Strong, "Engineers in Korea—Operation Shoestring," pp. 291–92.

labor who quite willingly undertook the strenuous construction tasks. Whole communities along the railroad lines mobilized for the reconstruction work.[17] The use of indigenous labor was absolutely necessary, as the 3d TMRS and the 32d Engineer Construction Group remained very much understrength and short of heavy construction equipment.[18]

Methods and Tools

U.S. Army engineer construction battalions assigned the larger bridge repair jobs—such as the 800-foot-long and 120-foot-high Killachon Bridge—moved in with heavy equipment and materials such as pile drivers, cranes, bulldozers, air compressors, and steel beams. Assigned the smaller bridges, Koreans moved in with masses of men, women, and children; sandbags; and native tools. The Koreans made up for their lack of heavy equipment with human numbers and often with great ingenuity in their repair work. The practical knowledge of native engineers supplanted formal training and standard American engineering practices. U.S. Army transportation and engineer officers supervised the overall work but left the methods and organization up to KNR officials and contractors.[19]

The native officials overseeing the work developed various expedients to replace the damaged or destroyed components of bridges. They replaced destroyed stone and concrete bridge piers with huge pyramidal structures made out of sandbags—some piled seventy-five feet high. For bridge stringers, the Koreans used rails, interlocking eleven lengths of rail tied together with baling wire, as a substitute for one heavy I-beam. The temporary bridges built by these methods slid and dipped, sagged and swayed, even as the first test locomotive ran over them, but they held up for several months remarkably well under the heavy traffic. The Korean laborers built five of these sandbag-and-improvised-steel-beam bridges in one seven-day stretch. More permanent steel-and-piling bridges replaced the temporary bridges as soon as time allowed.[20]

The Koreans also built permanent concrete bridge piers rapidly without the benefit of a mixing plant and dump trucks. Long lines of laborers passed sand, gravel, and cement in sandbags. Hordes of men, women, and children shoveled the sand, gravel, and cement together into a trough, to which a

---

[17] United Nations Civil Assistance Command Korea, United Nations Civil Assistance and Economic Aid Korea (1 October 1951–30 June 1952), RG 338, NARA; 3d TMRS, Unit History and Activity Report (1–31 October 1951), RG 407, NARA; Westover, *Combat Support in Korea*, p. 64.

[18] Strong, "Engineers in Korea—Operation Shoestring," p. 336.

[19] Miller, "High Steel in Korea," p. 332; Slevin, "World's Biggest Traffic Department," p. 9; Paschal N. Strong, "The Korean Builder," *Military Engineer* (September–October 1951): 336.

[20] Carl R. Gray Jr., *Railroading in Eighteen Countries: The Story of American Railroad Men Serving in the Military Railway Service 1862–1953* (New York: Scribner's Sons, 1955), p. 312; Strong, "The Korean Builder," p. 336.

hydraulic expert poured water. The expert gauged the proper water content by gurgle. The wet concrete was then carried by an endless line of laborers, each carrying a box of about one cubic foot capacity on his back. With a loaded box, a laborer walked up the ramp, dumped the load of concrete, and walked down again. The dumping was done by trap door in the bottom of the box opened by a string. The steady stream of concrete created by this method reportedly equaled the mechanical mixers and conveyors used by U.S. Army engineers.[21] When they required pilings for bridge trestles, the Koreans drove piles without a mechanical pile driver using a high tripod, a rope, a pulley, and a concrete block. The concrete block, which served as the driver, was alternately lifted and dropped by about twenty laborers pulling on the rope and chanting a song for proper timing.[22]

Manual labor and simple tools moved whole bridge spans as well. The Japanese had built the Korean railroad system largely for military purposes, on double-track lines. Most tunnels had separate bores for each track, and there were single-track bridges in parallel rather than a double-track bridge. This mitigated damage from air attacks and allowed for quick restoration of railroad traffic after an aerial attack.[23] The Koreans took advantage of this feature to rebuild bridges. Where there were two parallel bridges and the span of one bridge was out while the corresponding one on the other bridge was intact, the good span was manhandled over to the other bridge. They accomplished this by building cribs of cross ties connecting the two bridges and manhandling the girders on rollers until the span moved to its new location. They launched long, heavy girders by the cantilever method, with the girder being rolled out from the adjacent span by block and tackle and Spanish windlass. The most complicated piece of equipment in these operations was a screw jack. These efforts quickly opened single-track lines and left the other track with missing spans for later restoration as quickly as time and material allowed.[24]

The Koreans also proved very adept at salvaging steel from damaged bridges for reuse. When they needed reinforcing steel for a new concrete structure replacing a demolished one, they chipped the concrete away from the steel in the demolished bridge and used it for the new job.[25] All these jobs were undertaken with simple tools in lieu of mechanized equipment. Without sawmills, they split planks off logs using a sledgehammer and a steel wedge. This produced rough and unfinished planks of more or less uniform size. The hand saws that the Koreans used, based on Japanese practice, had crosscut teeth on one edge and rip-cut on the other, making them more efficient tools

---

[21] Huston, *Guns and Butter, Powder and Rice*, pp. 287–88; Strong, "The Korean Builder," pp. 337–38.

[22] Strong, "The Korean Builder," p. 338.

[23] Ibid.; Van Fleet, *Rail Transport and the Winning of Wars*, p. 33.

[24] Strong, "The Korean Builder," p. 338.

[25] Ibid.

than American single-purpose saws. They crushed stone for railroad roadbed ballast manually, using nothing more than sledgehammers.

An interesting tool used for placing and grading ballast was the three-man shovel. Working in unison, the chief operator thrust the shovel, attached to two lines, while two assistants pulled the lines. An American officer described it as "poetry in motion."[26] Using such tools, Korean track repair gangs worked with seemingly incredible speed; they repaired as many sections of track in three or four days as U.S. Army transportation and engineer personnel could repair in ten days using far more equipment.[27] Ironically, a few miles to the north of the battle lines, hosts of other Koreans and Chinese labored in railroad repair teams that undoubtedly employed the same techniques to keep enemy logistic lines open in the face of intense UNC aerial bombardment, naval gunfire, and guerrilla and commando raids. For this task, the Democratic People's Republic of Korea Railway Recovery Bureau had two construction brigades, and the Communist Chinese forces committed at least two railway construction divisions. By July 1953, as many as 40,000 Chinese railway construction troops were reportedly in North Korea.[28]

Needless to say, the expedient nature of repairs to track and bridges did affect railway operations. Derailments were a constant problem for the 3d TMRS and KNR in autumn 1950. The hastily restored track was subject to heavy rail traffic, and the situation was further compounded by the worn-out condition of the rolling stock. Six derailments occurred in one day, which completely overwhelmed the only heavy wrecking crane available on the entire railroad system. Again, indigenous skills and manpower proved of great use, as experienced KNR personnel cleared tracks and re-railed rolling stock with manual labor, readily available materials, and simple tools.[29] According to Capt. Max Brown of the 714th Transportation Railway Operating Battalion, "I marvelled at the ingenuity of the Koreans as they put freight cars onto the rail with little or no equipment."[30]

An Assessment

The contributions of indigenous labor clearly benefited KNR operations during the critical reconstruction periods in autumn 1950 and spring 1951, when army engineers and transportation corps personnel were hard-pressed to rebuild railroad lines rapidly enough to meet the support needs of advancing EUSA combat forces. According to Lt. Col. Howard W. Martens, assistant general manager for engineering, 3d TMRS, in this regard, "You can't give too much

[26] Ibid.

[27] Westover, *Combat Support in Korea*, p. 63.

[28] Charles R. Shrader, *Communist Logistics in the Korean War* (Westport, Conn.: Greenwood Press, 1995), p. 119.

[29] Westover, *Combat Support in Korea*, p. 66.

[30] Ibid.

credit to the Koreans."[31] Korean ingenuity and quick work in difficult bridge repair problems vividly impressed Colonel Martens.[32] Indigenous organization and methods in the bridge repair efforts also impressed Col. Paschal N. Strong, the EUSA engineer officer, who noted that

> the existence of the surprisingly efficient organization known as the Korean National Railroad was of immeasurable assistance in repairing railway lines and strengthening damaged railway bridges. True, their method of strengthening damaged trusses made the American engineer shudder, but not one of their crazily repaired structures has yet failed.[33]

Colonel Strong felt that the Army engineers could effectively learn from their experiences working with Korean builders. He observed that American engineers had become so accustomed to using mechanized equipment in construction projects that they forgot what could be done by hand alone. According to Strong, "With unlimited cheap labor around him, he [the engineer] is often unable to visualize ways and means to use that labor effectively in the absence of his favourite equipment."[34] The EUSA engineer officer felt that it would be beneficial for military engineers to study indigenous construction practices, as it would be useful for engineering operations in future overseas contingencies. Colonel Strong felt that vital construction work in a theater of operations could be readily accomplished without waiting for additional heavy equipment and material to arrive from the Zone of Interior.[35] While American engineer and transportation officers were favorably impressed with the ingenuity and alacrity of indigenous labor, their Korean counterparts were reportedly equally impressed with the speed and mechanized methods with which the American engineers reconstructed various long and high-level bridges.[36]

Conclusions

Indigenous labor certainly provided valuable assistance in rebuilding critically important railroad lines, as EUSA transportation and engineer units were understrength in personnel and short of heavy equipment and building material. Seemingly ancient practices helped to sustain a modern army with very heavy logistic needs. An existing local organization, the KNR, was of central importance in organizing and providing skilled labor for reconstruction work. This work took place with the hazards and urgency of wartime; accidents cost

[31] Slevin, "World's Biggest Traffic Department," p. 9.
[32] Ibid.
[33] Strong, "Engineers in Korea—Operation "Shoestring," p. 14.
[34] Strong, "The Korean Builder," p. 338.
[35] Ibid.
[36] 3d TMRS, Unit History and Activity Report (1–31 March 1951).

numerous lives during the rebuilding of bridges and the clearing of obstacles on railroad lines, which occasionally contained unexploded ordnance.[37] Why this work was done without coercion or hesitation was perhaps due, in part, to traditional Confucian values, a sense of parochial pride, patriotism, and genuine perception of an enemy threat. On a more basic level, people simply needed to eat. The agricultural life of the country was disrupted by the fighting and by the foodstuffs looted by Communist forces. The ROK was short of food, thus a hot noon meal was often more of motivating factor for labor than wages.[38] The efforts of the indigenous labor force served the EUSA very well. Judging from the comments of EUSA engineer and transportation officers, the contributions of native labor greatly assisted in restoring critically important railroad lines to service in autumn 1950. General James A. Van Fleet, EUSA commander, was "surprised" and "delighted" at the speed at which railroad lines were restored to service in April 1951.[39] Perhaps mechanized equipment might have been more efficient and American engineering practices more conventional, nevertheless, the often ingenious engineering methods and diligent work of native labor completed urgent reconstruction tasks in a timely fashion when military personnel and modern construction equipment were not available.

---

[37] Atkins and Myers, "The World's Worst Railroad Headache," p. 126; Kyotongbu, *Transportation of Korea 1957*, p. 43.

[38] Crump Garvin, "Pitfalls in Logistics Planning," *Military Review* (April 1962): 7.

[39] Van Fleet, *Rail Transport and the Winning of Wars*, p. 49.

PART THREE

U.S. Counterinsurgency Operations

The Victorio Campaign: Hunting Down an Elusive Enemy

Kendall D. Gott

The United States Army has often conducted operations in inhospitable climates and rugged terrain against an elusive and determined foe. Emilio Aguinaldo of the Philippines, Pancho Villa of Mexico, and, in recent times, the Muslim terrorist Osama Bin Laden faced the superior weaponry and materiel of the United States Army and put up a persistent and often effective struggle. Operations against these men were costly in manpower, bitterly frustrating, and took months of hard campaigning. The Army also faced extreme public scrutiny and a hostile press. The story of the Victorio campaign presents direct and relevant lessons for today's leaders.

Almost from its inception, the United States government sought to separate the growing number of settlers from the indigenous peoples by clearly defining tribal lands and keeping the native tribes within them. This goal evolved into a reservation system that encouraged the inhabitants to plant crops and live in settlements, thus giving up the old ways of hunting and gathering. The Bureau of Indian Affairs within the Department of the Interior administered the reservations. Unfortunately, the bureau was permeated by corruption and mismanagement.[1] With full stomachs and warm blankets, the American Indians might have become resigned to their new lives on the reservations, but too often rations were short or foul, and government corruption and inefficiency deprived them of promised blankets, clothing, and shelter. This situation left hunting or stealing as the only methods of survival; however, the reservations did not provide enough land for successful hunting. Although most of the tribes confined themselves to the reservations peaceably, bands of warriors often left these tracts and attempted to return to their traditional lands and old ways of life or, in some cases, simply to raid and pillage. The years 1879–1880 were particularly tense in the southwestern United States, as various bands simultaneously sallied out of their reservations. The U.S. Army was engaged almost everywhere across the vast frontier, protecting civilians and pack trains from attack or hunting down the wayward raiders.

The Army in the field during this period had no formalized doctrine for fighting Indians but adapted to the situation. The common tactic was an

[1] Robert M. Kvasnicka and Herman J. Viola, *The Commissioners of Indian Affairs, 1824–1977* (Lincoln: University of Nebraska Press, 1979), pp. 4–5.

offensive strategy that called for a drive into hostile territory against known native settlements, forcing the tribe to do battle or lose its food supply. Another method was the conduct of a relentless pursuit. Even if unable to catch its prey, a unit could, in theory, wear down an Indian force, compelling the Indians to leave the area. Defeating the Indians often became a matter of locating their camps and attacking them by surprise. This was achieved by adapting the standard use of night marches and dawn raids, catching the Indians asleep in camp. Experienced commanders also used deception, such as leaving campfires burning at night after the troops had moved or hanging back during a pursuit to lull the hostiles into a false sense of security.[2]

The backbone of the Army effort to secure the frontier was the cavalry, but it had its limitations. Perhaps the most restrictive was its horses, which were generally bigger and stronger than their native prairie counterparts but were accustomed to a diet of grain fodder, which required substantial logistic support. While on campaign, units generally hauled supplies by wagon trains, although some preferred pack mules. The wagons slowed columns down, while the mules traversed the rugged terrain with ease but could not carry loads as efficiently. Either method was extremely vulnerable to interdiction by the hit-and-run raids of the fast-riding Indian warriors. One of the Army's innovators during this period, Maj. Gen. George Crook, a veteran corps commander of the Civil War and commander of the Department of Arizona from 1871 to 1875, studied the Apache way of war and decided to implement major changes. He determined that the best way to fight them was to copy their techniques of rapid movement. He instituted the widespread use of mule trains while stripping the baggage trains and individual soldiers of excess weight and training his soldiers in the techniques of long and mobile campaigns.

In another pivotal development, the Army enlisted Indians into the service for six-month tours as scouts. It easily recruited scouts by going to a rival tribe that harbored ancient hostilities or to competing factions within the hostile tribes who felt that the active renegades constituted a long-term threat to future peace and prosperity. These scouts ably tracked even the most elusive warrior party and proved invaluable as guides in finding water, provisions, and trails. They also served occasionally as couriers and engaged in actual combat. Organized into companies of twenty-six men and led by white officers, the Indian scouts were attached to a specific command for the duration of a campaign, receiving their orders from the senior officer present. Often, these companies were fragmented by assigning small squads to various detached commands. The Indian scouts earned a reputation for dependability and valor throughout the Indian wars. Without their efforts, the defeat of the hostiles would have taken far longer and cost many more lives.[3]

---

[2] Andrew J. Birtle, *U.S. Army Counterinsurgency and Contingency Operations Doctrine, 1860–1941* (Washington, D.C.: U.S. Army Center of Military History, 1998), p. 73.

[3] Joseph A. Stout Jr., *Apache Lightning: The Last Great Battles of the Ojo Calientes* (New York: Oxford University Press, 1974), p. 34. George Crook was born on 8 September 1828

The tribes of the Apache were a tough enemy. The desert afforded little means for a sedentary agrarian society, and the Apache had maintained a mostly nomadic culture, with an economy and political system based on raiding and plundering. Not surprisingly, this kept them at war with their neighbors through the centuries. The constant warfare and ability to live in the mountains made the Apache renowned for their fighting prowess and an astonishing ability to endure pain and hardship. A tight-knit clan, these people had a callous disregard for outsiders regardless of the hue of their skin, and looked upon them all as essentially something less than human. The tribal social hierarchy was structured by a status measured on success in battle and how much plunder a warrior acquired in the raids. The great chiefs attained and held their power by force of personality backed by their continued success. This hold on power was precarious, however, as the ambitious braves, who had been taught from early childhood to hunt, track, ride, and fight, vied for position and status for themselves.[4]

A product of this warrior culture, Victorio lived through the events that saw his people fall from a great tribe to settlement on reservations. Very little is known of his early life, but he was probably born in the Black Mountain Range of New Mexico around 1820 and reared as a member of the Eastern Chiricahuas, often referred to as the Mimbres Apache. Victorio rapidly rose in influence and emerged as a full tribal leader, forming a group of some four hundred fanatically devoted warriors. Using the Warm Springs reservation in New Mexico as a base of operations, these warriors mercilessly raided their enemies, taking what they wanted and destroying what they did not.[5] Mexico and the United States exchanged mutual protests for failing to stop Apache raiders from crossing the border and giving them sanctuary upon their return. American settlers also placed immense pressure on their representatives in Washington to halt the local Apache raiding and killing.

Victorio received no punishment for his depredations, but public outcry grew. Tension on the reservation grew as well, as subdued Apache clans feared reprisals for his activities. When Victorio saw an approaching column

---

near Dayton, Ohio; graduated from West Point in 1852; and finished the Civil War in command of a cavalry division. Although he did not invent the idea of using Indian scouts, Crook made full use of them. By 1886, the Indian scouts were issued regulation uniforms with distinctive insignia. See also Douglas C. McChristian, "Pueblo Scouts in the Victorio Campaign," in The Military Frontier, A Symposium to Honor Don Russell, Buffalo Bill Historical Center, Cody, Wyo., 2–4 May 1986; copy held at Fort Davis National Historic Site.

[4] Stout, *Apache Lightning,* p. 4. American Indian tribes were exceptionally diverse in customs, reflecting the regions in which they lived. The Apache were perhaps the extreme of the warrior culture. Especially while they were on the warpath, the Apache paid little heed to the needs of their animals, reasoning that if they broke down, they would be the next meal, and the next target would provide replacements.

[5] Ibid., pp. 76–77. Victorio was known by many names in his life, including Victoria, Vitoria, Vittorio, Beduiat, Bidu-ya, Lucero, Light, and Laceres. His sister was the famous female warrior Lozen. Members of the tribe were fanatically loyal to him.

of soldiers, he and his followers hastily left the reservation, believing he would be tried for old murders and horse stealing and sent to prison or to exile in Florida. Preferring death to either, he fled on 21 August 1879 with about eighty warriors, along with their wives and children.[6]

Once off the reservation, Victorio's band attacked an Army outpost near Warm Springs, killing five soldiers and three civilians and making off with the sixty-eight horses and mules of Company E, 9th Cavalry (*Map 1*). A few days later, Victorio struck another outpost some twenty miles south of the reservation. His band killed ten soldiers and captured all of the livestock. In response, the 9th and 10th Cavalry regiments deployed to the field. The Army also consolidated additional troops from scattered posts throughout the district to meet this threat and brought in Lt. Charles Gatewood and his Apache scouts from Arizona. Cavalry detachments, led for the most part by junior officers, deployed across New Mexico with orders to find Victorio's elusive band.[7]

Victorio's force varied in size, starting with eighty warriors and growing to no more than three hundred warriors at any given time. Including women and children, his band never exceeded 450 souls. As Victorio's success and fame grew, young warriors flocked to him or at least emulated him by raiding in their own local areas. These scattered incidents inflated Victorio's actual strength and gave observers the impression that he was able to move fantastic distances over a short period. When faced with strong opposition, Victorio's band would often split into small groups and cross into the safety of Chihuahua, Mexico, before they could be intercepted.

Victorio sustained his operations over several months, primarily by acquiring what he needed from his defeated foes. His men took arms and ammunition from dead opponents, from mercantile traders, and from Mexican sheepherders whom they happened upon and killed. The Apache often rode their horses to death, then ate the dead mounts and simply stole more to carry on the fight. Since everything Victorio used was readily available, he could travel light and fast. The one exception to the rule—and his primary logistical concern—was water. In the mountains and desert plains, even the hardy Apache could not carry large enough amounts of the essential water. With this constraint, Victorio had to plan his operations around the few sources of water in the region. The few times the U.S. Cavalry engaged Victorio in open battle in the first nine months of his departure from the reservation was usually when his band stopped and set up camp around a watering hole. But battles, such

---

[6] Dan L. Thrapp, *Victorio and the Mimbres Apache* (Norman: Oklahoma University Press, 1974), pp. 219–20. This book is currently the definitive work on Victorio and the campaign. The actual incident that prompted Victorio to leave the Warm Springs reservation varies with a wide number of sources. Whatever that incident was, he was not happy there, and it took little provocation.

[7] Ibid., pp. 236–37; William H. Leckie, *The Buffalo Soldiers: A Narrative of the Negro Cavalry in the West* (Norman: University of Oklahoma Press, 1967), pp. 210–11.

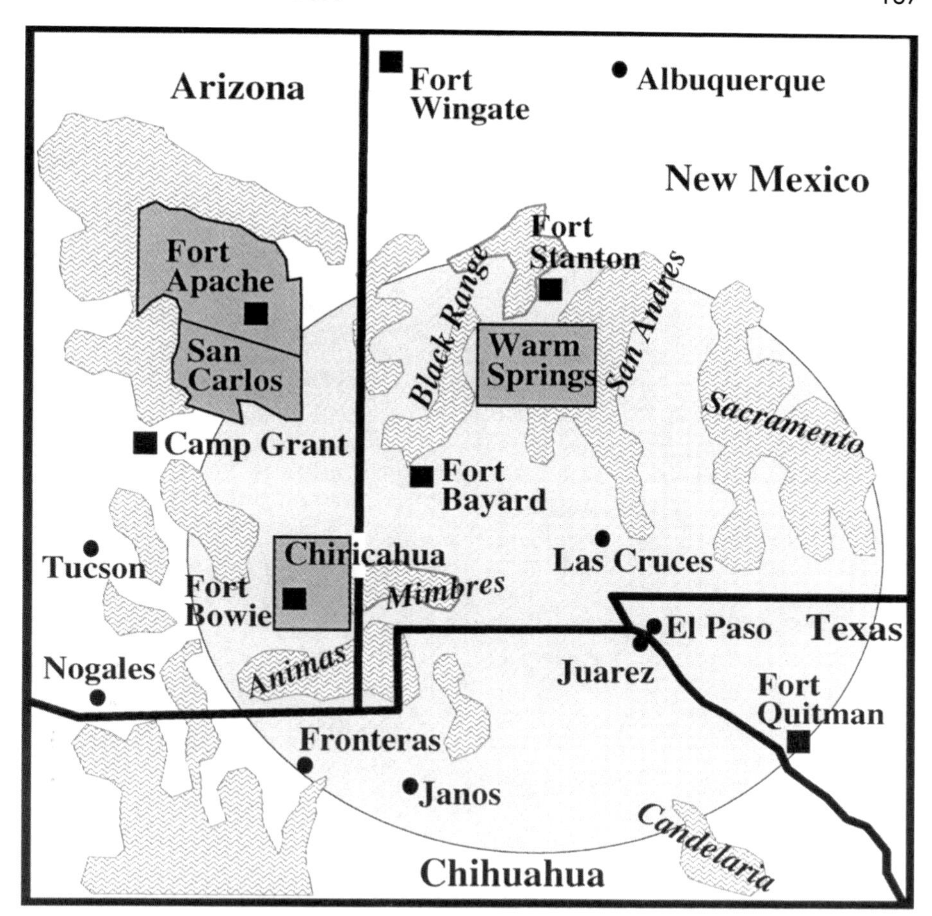

Map 1—The Area of Operations, 1879–1880

as those at Las Animas and Hembrillo canyons, were indecisive, as Victorio eluded his foes and slipped away to Mexico, only to return again.

The Fights at Tinaja de Las Palmas and Rattlesnake Springs

In the summer of 1880, Col. Benjamin Grierson, commanding the 10th Cavalry regiment, surmised that Victorio meant to reenter the United States with the likely objective to head straight for the Mescalero country of southern New Mexico in search of supplies and new recruits. Grierson strengthened the subposts along the Rio Grande at Viejo Pass, Eagle Springs, and old Fort Quitman, which had been abandoned as a permanent post three years earlier. Yet concentrating combat power was not enough. It was critical to determine Victorio's location in order to apply it. On 28 July, Grierson learned from his scouts, who had crossed the border into Mexico, that Victorio was indeed headed north toward the Rio Grande with 150 warriors. Knowing the area well, Grierson guessed Victorio would have to stop for water at Tinaja de Las Palmas, which was presently unguarded since it held water only after a rain and was thus a generally unreliable source. The spotting of a lone Apache reconnoitering in the area strengthened this assessment. Grierson ordered the companies and detachments from the various posts to converge at this key water hole.[8]

Colonel Grierson's initial detachment reached Tinaja de Las Palmas at the foot of Rocky Ridge and entrenched just short of the crest. He had just twenty-three men to hold the three rock redoubts erected to guard the water hole. Grierson needed reinforcements and sent two riders at a gallop to bring up whatever forces they could find. At 0730, Pueblo Indian scouts reported that Victorio had camped in a canyon only ten miles to the south and was preparing to move. Time was running out for Grierson and his small band of Buffalo soldiers (*Map 2*).

At 0900 on 30 July, the weary Apache approached the water hole and quickly spotted the blue-clad troopers. Victorio sent his men to the east to bypass this position, but Grierson would not allow him to get away that easily and ordered Lt. Walter Finley to charge forward with ten men. These soldiers advanced into a classic ambush—Victorio had deliberately exposed part of his force as bait to lure his foe into a killing zone. Finley's force was out of range of support from the small forts and on its own. The Apache took cover among the rocks and returned fire in a skirmish that lasted over an hour. They presented the soldiers with a grave danger of encirclement and annihilation in their advanced position. Desperate, Finley ordered a charge to break the Apache siege. Hearing the sound of battle in the mountains some distance away, Capt. Charles Viele, with Companies C and G, charged down the road from Eagle Springs and joined the fight at this critical time. However, in the smoke and dust of battle, they mistook Finley's detachment for hostiles and opened fire, forcing it to withdraw from its advanced positions back to Grierson's troopers covering the water hole. Taking advantage of the confusion, Victorio,

[8] Leckie, *The Buffalo Soldiers,* p. 225; Douglas McChristian, "Grierson's Fight at Tinaja de Las Palmas: An Episode in the Victorio Campaign," *Red River Valley Historical Review* 7, no. 1 (Winter 1982): 54–59. This fight is also known as the Battle of Rocky Ridge. See also Edward L. N. Glass, *The History of the Tenth Cavalry: 1866–1921* (Fort Collins, Colo.: Old Army Press, 1972), pp. 12–23, for a detailed history of the regiment prior to the campaign. Benjamin H. Grierson was born on 8 July 1826 in Pittsburgh, Pennsylvania, and was a music teacher and storekeeper. During the Civil War, he was a major in the 6th Illinois Cavalry and later was promoted to colonel. Grierson pursued Van Dorn after the Holly Springs raid and led his own famous raid through Tennessee and Mississippi in June 1863. He was appointed a brigadier general and commanded a cavalry division in the Army of the Mississippi. On the frontier, he later commanded the Department of Arizona and the Districts of New Mexico and Indian Territory. Grierson retired in 1890 and died on 1 September 1911 in Omena, Michigan.

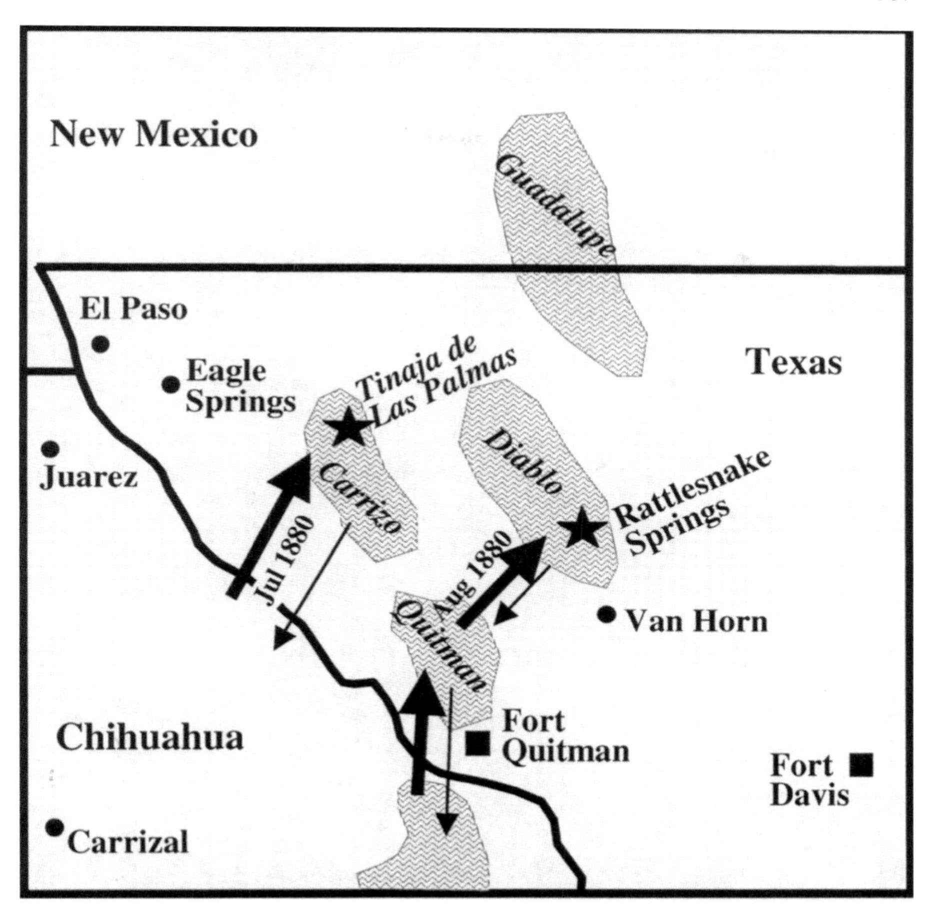

Map 2—Victorio's Last Two Raids, Summer 1880

who apparently underestimated the numbers of the force he faced, ordered a counterattack. The screaming Apache rose up and rushed forward. Once out from behind their cover, Victorio's men made easy targets, and the cavalry stopped the attack in its tracks. A short reprieve followed, but Victorio's braves regrouped and renewed the fight. It took another hour, but Captain Viele finally fought his way through to Colonel Grierson. Seeking to avoid battle against diminishing odds, Victorio again tried to bypass the cavalrymen and head north, but Grierson ordered another detachment forward, which cut off the warriors and forced them to turn back.[9]

[9] McChristian, "Grierson's Fight," pp. 60–61; Robert K. Grierson, Journal Kept on the Victorio Campaign in 1880, p. 24. Copy held at Fort Davis National Historic Site (National Park Service), Fort Davis, Texas, and cited with permission. Hereafter cited as Robert's Journal,

By 1230, the outnumbered Mimbres broke contact and scattered southward toward the Rio Grande and Mexico beyond. The cavalry suffered surprisingly light losses: one man killed and one wounded in the four-hour fight. Victorio lost approximately seven men killed from his force of about one hundred braves and a large number of wounded. This skirmish forced Victorio to retreat, but Grierson knew his adversary would soon return. No one realized it at the time, but this tactical defeat meant the beginning of the end for Victorio.[10]

Colonel Grierson would not have to wait long for Victorio's return. The word that Victorio was in motion again arrived in a message from Col. Adolpho Valle of the nearby Mexican forces in the final week of July. U.S. troopers spotted Victorio's advance guard, and Grierson ordered his scattered detachments to converge on Eagle Springs. A cavalry detachment made brief contact with Victorio's main force on 3 August in a surprise engagement, and other patrols confirmed that Victorio and his followers were headed north toward the Guadalupe Mountains. Grierson ordered his command in that direction in an attempt to block the Apache. He hoped to set up an ambush in Bass Canyon, near the town of Van Horn, and amended his orders to his scattered companies to converge there.[11]

On 5 August, Grierson's 10th Cavalry raced northward some sixty-eight miles in twenty-four hours on the east side of the mountains, shielding his command from observation, with the objective of reaching the watering hole at Rattlesnake Springs, one of but two permanent water sources in the region, before Victorio arrived. (The other permanent water hole was located at Sulphur Springs, some fifty miles away.) It simply became a race to reach the vital source of water first. Interestingly, Grierson countermanded his orders to leave the wagons behind. Instead, the cavalry would use the lumbering vehicles as far as possible. Although he left no record to explain this decision, it meant that Grierson would have a reliable means of supply and would not leave the valuable trains vulnerable in the open terrain.

Traversing the desert and mountains for days wore down the cavalrymen and their mounts, but Grierson and the advance elements of the 10th Cavalry marched another sixty-five miles in under twenty-one hours and reached Rattlesnake Springs, arriving just ahead of Victorio. The large quantities of forage and water carried by the wagons made this rapid advance possible, a rare example of their use actually speeding an advance. After posting his men

---

this is a very colorful account of the campaign, in which the young Grierson made daily entries while accompanying his father on the campaign.

[10] Leckie, *The Buffalo Soldiers,* p. 225; Rpt, Col B. H. Grierson to Assistant Adjutant General, Department of Texas, 20 Sep 1880, doc. G, in *Report of the General of the Army, Annual Report of the Secretary of War* (Washington, D.C.: Government Printing Office, 1880), vol. 1, paras. 16–20. A copy is held by the Fort Davis National Historic Site (National Park Service) Fort Davis, Texas. Hereafter cited as Grierson's Report, this text is used extensively in this work in detailing the chronology of the battles of the 10th Cavalry and Grierson's assessments and rationale. See also McChristian, "Grierson's Fight," p. 61.

[11] Grierson's Report, vol. 1, paras. 16–20; Robert's Journal, pp. 32–34.

in the rocks to cover the water source, Grierson awaited reinforcements. Soon Captain Carpenter and two more companies joined him and were posted a short distance south of the springs in support.

At 1400 on 6 August, the Apache slowly made their way down Rattlesnake Canyon toward the springs, quite unaware of the ambush laid for them. They too had ridden hard, and both warriors and horses needed water. Victorio was tactically off-balance and apparently did not have an advance guard. Just seconds before the signal was given to the cavalrymen to commence fire, Victorio sensed danger and halted his men. With hostiles in their sights who were about to bolt, the troopers did not wait; they opened fire on their own initiative, and under a hail of lead the Apache scattered and withdrew out of range.

But Victorio's people needed water, and, believing there were only a few soldiers present, they regrouped and attacked immediately. However, Grierson had posted two companies in an overwatch position covering the water, and a few massed volleys from their carbines sent the hostiles scattering back into the canyon. Stunned by the presence of such a strong force, but in desperate need of water, Victorio repeatedly charged the troopers in attempts to secure the spring. When the last such attempt failed near nightfall, Victorio and his followers withdrew southwest deeper into the mountains. The cavalrymen were in hot pursuit, but darkness finally halted Grierson's men. Victorio was now critically short of food and water and facing increasing numbers of soldiers, rangers, and armed citizens. His people were hardy but not indestructible. With resupply uncertain and faced by such strength, Victorio slipped his band back across the border into Mexico to avoid pursuit. His people had lost over thirty braves killed and fifty wounded, and almost all their horses.[12]

The defeat substantially weakened Victorio, and Grierson now had most of his command assembled. Giving his enemy no rest, Grierson organized his force into three squadrons of two companies each and sent them to comb the mountains for signs of the hostiles. He also maintained guards on all of the known water sources. Victorio's braves, women, and children had spent a year on the run, and the harsh conditions and deprivation had begun to tell. The constant skirmishing with the American and Mexican forces had worn down the people and animals. Indian scouts reported that Victorio had crossed the border on 12 August, but it was apparent that the numbers of wounded warriors and broken-down stock slowed his march. This trail was to be the last evidence seen of Victorio in the United States. However, Victorio still survived, and while no one could know when he would return, everyone knew that the raids and killing would continue as long as he lived.

Victorio and his band crossed the border, but the Mexican forces avoided a decisive battle and, in effect, allowed them safe passage. The Apache chief holed up in the mountains deep into Mexico. The rugged mountains provided safety but held no game and could not support his people for long. Effectively

---

[12] Thrapp, *Victorio and the Mimbres Apache*, pp. 238–40.

blocked from returning to New Mexico, Victorio lead his people southeast and farther into Mexico to a mountain range known as Tres Castillos (Three Peaks) in the hope of finding a sanctuary. These unimposing mountains were essentially no more than three mounds of rock in a vast open desert, with little water, sparse grass for the animals, and no means of escape except by crossing the open desert once again. Instead of a place of refuge, Victorio had found a final trap.

The Mexican forces in pursuit, under the command of Col. Joaquin Terrazas, had over 350 men under arms with ample supplies and ammunition. Terrazas' scouts soon found the trails leading to Victorio's camp, and the Mexican force quickly converged on that location and surrounded it on 14 October (*Map 3*). For twenty-four hours, the adversaries waged a bitter and bloody battle among the rocks. The Apache were soon almost out of ammunition and were reduced to throwing stones. A Tarahumari sharpshooter felled Victorio himself with a miraculously long shot, although Apache legend persists that "Old Vic" took his own life with his knife to prevent capture.[13] When the fight ended, Victorio and eighty-six warriors lay dead, and eighty-nine women and children became prisoners. The Mexicans held them in Chihuahua City for the next several years. Only a handful of warriors escaped, scattering and eventually making their way back to the reservations in New Mexico and Arizona.

Victorio's demise seriously weakened the resistance of the Mimbres Apache. The number in any given raider band became generally no more than ten warriors, with up to thirty being the rare exception. Forays by notable Apache such as Nana and Geronimo were brief and small affairs when compared to the size of Victorio's force and the destruction it had caused. By 1886, the Army had finally pacified the Apache, and peace came to the Southwest.

The Final Analysis

The Victorio campaign illustrated how a small band of dedicated and hardy people can fight for an extensive time against an overwhelming force. By knowing the terrain intimately, Victorio usually stayed one step ahead of his pursuers, savagely striking at will and disappearing as a phantom into the mountains. As his people could endure extreme hardship and Victorio knew the location of the few sources of water, the renegade Apache formed a constant and deadly threat to the settlers on both sides of the border.

Victorio and his supporters maintained that had the Mimbres Apache been allowed to stay at the reservation of their choice, they would have been peaceful and content. That may or may not have been true, but the issue

---

[13] Ibid., pp. 301–04, is a detailed account of the final battle. Various accounts of Victorio's death were told after the fight, including one fanciful version in which he was mounted on a fine white horse. Victorio wore no distinctive garments and was probably identified only after the battle.

Map 3—Victorio's End, October 1880

became irrelevant upon their final departure from the reservation. Victorio and his band were convinced that there could be no peace with the American government, and after a series of murders and stealing, they were absolutely correct. The public would not tolerate their simple return to the reservation to take up the plow. This left but two outcomes for Victorio: victory or death. The inexorable flow of Western civilization across the frontier guaranteed the final outcome in the long term. The Apache would adapt to the reservation or they would die trying to maintain their old way of life.

Three key elements brought about the defeat of Victorio by the U.S. Cavalry. First and foremost was the use of indigenous personnel as trackers and guides. The Indian scouts reliably identified Victorio's whereabouts and predicted his intentions. They also penetrated into Mexico and gathered intelligence almost at will. If they were caught by authorities, the U.S. government could easily disavow them. Without the Indian scouts, the cavalry would have been limited to pursuit operations against a foe that could afford to drive its animals to death and then simply steal more to continue on.

The second key element was Colonel Grierson's decision to forgo offensive search-and-destroy operations, which had no telling effect on the enemy's logistics and support, and, instead, to concentrate on the foe's logistical Achilles heel: the need for the few sources of water available in the desert. By placing guard forces to cover the few sources of water in his department, he denied the enemy a resource essential to successful operations. This new tactic forced Victorio to come to a location where the cavalrymen would enjoy the advantages of defending among the rocks. Being deprived of water and lacking the firepower to take it, Victorio was forced to withdraw in an increasingly desperate search for it. Given Grierson's control over the available water sources, Victorio had few choices. Faced by a strong cavalry screen and short on water, he returned to Mexico to meet his fate weeks later.

Last, the determination of Grierson to pursue Victorio's band proved decisive. The innovations in the use of pack mules and wagons made this possible. Constantly on the run, the Apache found it increasingly difficult to rest and regroup. This strategy required Grierson to keep his men in the field and forced immense costs in supplies and horses. He also had to contend with an increasingly frustrated public and the press. However, over time, patience and arduous efforts wore Victorio's band down to a point that it withdrew to Mexico.

The lessons of the Victorio campaign are as relevant today as they were in the nineteenth century. Commanders must understand the enemy's methods of operation and exploit his weaknesses as Grierson did by depriving Victorio of essential water. Commanders and staffs must also look beyond their formal training in devising flexible tactics and strategy and in preparing their units for sustained operations that might last for months. By following these tenets, so aptly illustrated in the Victorio campaign, elusive enemies can be tracked, worn down, and defeated.

Without the Need of a Single American Rifleman: James Van Fleet and His Lessons Learned as Commander of the Joint United States Military Advisory and Planning Group During the Greek Civil War, 1948–1949

Robert M. Mages

Lt. Col. Paul J. Ciesinski was determined to take an active part in America's Global War on Terrorism. Not content with the domestic security tasks assigned to his unit in the Connecticut National Guard, Ciesinski transferred to the U.S. Army Reserve's 80th Infantry Division (Institutional Training) and volunteered to lead a Military Transition Team (MiTT) assigned to advise a brigade of the Iraqi Army. A dedicated patriot and student of military history, Ciesinski understood the importance of this mission and looked forward to building this new force. Unfortunately, his first challenge "was recovering from the 'poorly planned and insufficient' predeployment training" he and his soldiers received.[1] None of the instructors or training cadre charged with preparing the MiTT seemed to understand the nature of the task that awaited Ciesinski's men in Iraq. He found the U.S. Army's failure to provide these soldiers with useful doctrine on military assistance and adviser operations the most disappointing of all.

> We were really astounded by the lack of doctrine. At least in 2005 and into 2006 there was zero doctrine on how to advise. This astonished me because our Army—not just the special operations community—from the early 1960s to 1973 advised the South Vietnamese Army. There should have been a lot of lessons learned. I couldn't believe there wasn't any doctrine. There wasn't anybody around to tell us what an advisor even does. That was one of the classes I gave to my subordinates, on what advisors were supposed to do—and I based that on historical example and just common sense. There was nobody there to tell us what advisors do. The impression I had was, "Just go to Iraq and tell them what we think is right." The Army itself had no doctrine so it appeared our trainers didn't think it was important to train us on that.[2]

[1] Interv with Lt Col Paul Ciesinski, 2007, p. 2, U.S. Army Combat Studies Institute, Fort Leavenworth, Kans.

[2] Ibid., p. 7.

This failure of the Army's institutional memory is puzzling and regrettable. The advisory efforts carried out by the United States Army during the Cold War have been well documented by Army historians. Anyone charged with writing a new doctrine on advising foreign armies will find a rich historical record to draw on. As we work to rebuild the forces of our allies in the Middle East and central Asia, we must continue to review this record for lessons and general principles that we can impart to advisers like Colonel Ciesinski.

This paper will examine the American advisory effort in support of the Greek National Army (GNA) during the Greek Civil War (1947–1949). It will place specific emphasis on the actions of the commander of that effort, Lt. Gen. James A. Van Fleet, in order to extract general principles that can benefit our ongoing attempts to help our allies build effective national armies in support of America's war on terror.

A Brief History of the American Advisory Effort

Greek nationalist and Communist forces had been locked in a cycle of escalating civil strife and bloodshed since the end of World War II. A full-scale civil war broke out after a plurality of the Greek people (68.9 percent voted for the return of the king) voted by plebiscite in September 1946 to establish a constitutional monarchy and a parliamentary democracy.

The Greek Communist Party (KKE) refused to participate in the government. It re-formed its guerrilla forces, rechristened them the Democratic Army of Greece (DAG), and established headquarters in the mountains of northern Greece on 28 October 1946. This rebel army consisted of approximately 22,000 insurgents, led by cadres hardened by years of conflict with Axis invaders and nationalist forces. Established in mountain strongholds throughout the country, the DAG unleashed a vicious guerrilla campaign that pushed the 120,000-man Greek National Army back into the towns and cities, cut communications across the country, and threatened to overwhelm the weak central government.[3]

The Soviet Union and Marshal Josip Broz Tito's Communist regime in Yugoslavia supported the KKE. They provided Soviet weapons and equipment to the insurgents and allowed the construction of base areas over the border in Albania and Yugoslavia. Only Great Britain was willing to stand with the Greek government. The British had long been involved in Greek affairs and had already fought a round with the insurgents during the winter of 1944–1945. However, the military and economic exhaustion of postwar Britain precluded them from providing the quantity of support needed to sustain the Greek government. In desperation, the Greeks turned to the Americans for assistance.

[3] Edgar O'Balance, *The Greek Civil War, 1944–49* (New York: Praeger, 1966), pp. 111–20.

In March 1947, the Greek government presented the American ambassador to Greece, Lincoln MacVeagh, with an urgent request for aid in its struggle to overcome a powerful and increasingly dangerous Communist insurgency.[4]

Help arrived in the form of a bill signed by President Harry S. Truman on 22 May 1947, pledging $300 million to the Greek cause and directing that an aid mission and military advisers be sent to the country. The Cold War had dawned, and the United States found itself engaged in a global conflict against the Soviet Union and its hostile, expansionist ideology. In response to this threat, the Truman administration developed a strategy to contain and combat this menace. In what would become known as the Truman Doctrine, the American president declared, "I believe that it must be the policy of the United States to support free peoples who are resisting attempted subjugation by armed minorities or by outside pressures." The Greek Civil War would be the first test of this global strategy.[5]

On 31 December 1947, the Joint Chiefs of Staff directed the creation of the Joint United States Military Advisory and Planning Group (JUSMAPG). They ordered the group "to assist the Greek Armed Forces in achieving internal security in Greece at the earliest possible date."[6] Maj. Gen. William G. Livesay commanded the group for the first few months, but he did not fare well. Prior to taking his post, General Livesay was told by General Lauris Norstad, director of the Plans and Operations Division of the War Department, that his mission was to train the Greeks in the proper use of American equipment and that matters of overall training and reorganization were the responsibility of the Greeks and their British allies.[7] He was not empowered to provide operational advice to the GNA and soon lost the confidence of many within the Greek government. At a meeting with Secretary of State George Marshall, Queen Frederica of Greece complained that her country needed "a fighting general," not a "supply sergeant." This judgment was not altogether fair to Livesay, but the queen managed to convince Marshall that he needed replacement. After conferring with Army Chief of Staff and General of the Army Dwight D. Eisenhower, Marshall selected General Van Fleet as the best candidate to take command of JUSMAPG.[8]

---

[4] John O. Iatrides, *MacVeagh Reports, Greece, 1933–1947* (Princeton, N.J.: Princeton University Press, 1930), pp. 714–20.

[5] American Mission for Aid to Greece, A Factual Survey Concerning the American Mission for Aid to Greece, American Mission for Aid to Greece, Athens, 1948 (hereafter cited as AMAG Survey), p. 18.

[6] Joint U.S. Military Advisory and Planning Group, Greece, History of JUSMAPG, 1949, Headquarters, JUSMAPG (hereafter cited as JUSMAPG History), p. 2.

[7] Maj Gen William G. Livesay, Personal Diary, 13 Jun 47, records his meeting with Gen Lauris Norstad and the general guidance provided by the War Department.

[8] James A. Van Fleet, "How We Won in Greece," 11 April 1967, speech to the Institute for Research in the Humanities and University Extension, University of Wisconsin, Madison, Wisc.

Van Fleet was a graduate of West Point and a thirty-year veteran of the U.S. Army. He had seen combat in both world wars and served with distinction in the European Theater of Operations (ETO) during World War II. Eisenhower directed Van Fleet to report to Marshall. After a brief interview, the secretary of state was satisfied that Van Fleet was the correct choice. He was promoted to lieutenant general and dispatched to Athens in February 1948.[9]

The decision was a fortunate one for both Greece and the U.S. Army. Van Fleet brought a warrior focus and a sense of urgency to the aid effort and the GNA. He was determined to wage an aggressive campaign against the insurgents while simultaneously implementing a rapid and comprehensive reorganization of the Greek Army. He directed the JUSMAPG to assist the Greeks in planning and executing a series of offensive operations designed to trap and destroy the insurgents in their mountain bases.[10]

Under Van Fleet's guidance, Greek forces worked to clear the areas around Athens and force the DAG to withdraw to its mountain sanctuaries on the Albanian border. The offensive proceeded slowly and produced an indecisive outcome. The GNA followed up these attacks in late June with Operation CROWN, a three-phase operation designed to destroy DAG forces in the Grammos Mountain region. Once again results were mixed. The poor operational security and shoddy execution of the GNA allowed the DAG to withdraw into Albania and reenter Greece around Mount Vitsi near the Yugoslav border. Vowing to give the enemy no rest and determined to destroy him before he could strengthen his mountain bastion, the GNA, under the direction of the Greek General Staff (GGS) with guidance from JUSMAPG, launched a series of attacks in the Vitsi area from August to October of 1948. These too achieved disappointing results. The national army failed to dislodge the insurgents from their defensive positions, and the fighting dissolved into a bitter stalemate. While the DAG emerged from these battles emboldened and stronger, they exposed the GNA as a timid and ineffective. The results on the battlefield did not seem commensurate with the resources expended, and doubts about the aid effort began to grow in Washington and Athens.[11]

Undeterred, Van Fleet worked throughout the winter to rebuild the GNA, using American training methods and weapons. Meanwhile, the KKE decided to change tactics. Growing personal and ideological differences provoked KKE leader Nikos Zahariadis to depose the Communist military commander Markos Vaphiadis and purge the senior ranks of the DAG. Vaphiadis had organized the DAG into company- and platoon-size detachments engaged in raids, ambush, and sabotage while avoiding direct combat with large GNA formations. Zahariadis converted the army into a more conventional force,

---

[9] R. Manning Ancell with Christine M. Miller, *The Biographical Dictionary of World War II Generals and Flag Officers; The U.S. Armed Forces* (Westport, Conn.: Greenwood Press, 1996), p. 330.

[10] JUSMAPG History, pp. 14–15.

[11] Ibid.

organized into brigades and divisions, prepared to take and hold ground in defiance of the national army. In order to boost recruiting in the border areas, he also embraced the cause of Macedonian autonomy. Both of these steps would have catastrophic consequences. The premature transition to positional warfare played into the hands of the reconstituted GNA, and the political shift on the Macedonian question infuriated Tito, forcing him to reconsider Yugoslavia's support for the KKE insurgency.[12]

In late April 1949, the reconstituted GNA began a series of operations to clear the insurgents from the mountain areas of central Greece. Instead of melting away and withdrawing north, the DAG commanders chose to fight it out. Practicing the tactics taught to it by Van Fleet's advisers and using its new American weapons to good effect, the GNA cleared the DAG from these regions, opening the avenues of approach to the insurgent base areas along the Albanian and Yugoslavian borders. Sensing that the tide of battle had turned and piqued over the KKE's position on Macedonia, Tito closed his border and ceased his support for the insurgency.[13]

Having seized the initiative, Van Fleet and the GGS ordered commencement of the final, decisive offensive. They launched Operation TORCH against the last Communist strongholds in the Vitsi and Grammos Mountain regions. Within a few weeks, the insurgents were routed and the remnants of the DAG withdrew into Albania, which disarmed and interned them.[14]

The victory of the GNA was also a triumph for JUSMAPG and its commander. Van Fleet's success was based upon the application of four basic principles:

- Work with and support elements in the indigenous armed forces and government that share American goals and objectives.
- Demand accountability.
- Build the indigenous army according to the requirements of the conflict.
- Remember that the war must be waged and won by the indigenous army and the government it defends.

Finding Friends

The political goal of the United States was to "encourage within the framework of the Greek Constitution and without direct intervention, the maximum unity among the different loyal elements of the Greek political world." The U.S. Embassy considered the Greek Populist-Liberal coalition government established in September 1947 and headed by the Liberal Prime

---

[12] Averoff-Tossizza Evangelos, *By Fire and Axe: The Communist Party and the Civil War in Greece, 1944–1949*, trans. Sarah Arnold Rigos (New York: Caratzas Bros., 1978), pp. 285–300.

[13] O'Balance, *The Greek Civil War*, pp. 185–95.

[14] Ibid.

Minister Themistocles Sophoulis as "an important step in the movement towards unity."[15] The embassy staff sought first to preserve, and later build on, that fragile unity in the hope that the Communist KKE could be persuaded to end the insurgency and embrace the Greek Constitution.[16] This strategy often caused friction between the embassy staff and JUSMAPG.

Van Fleet did his best to stay aloof from Greek political infighting, but it soon became clear that elements within the Greek government were not altogether pleased by American intervention. Even supportive factions disagreed over matters of military strategy and with the decisions made regarding the training, equipment, and composition of the GNA. Well aware of these currents, Sophoulis counseled Van Fleet on how best to avoid Greek political entanglements.

> General please give our Greeks proper weapons and training and develop some operations for them and see that they carry them out. I look upon you as our savior and will support you in any way you wish, to save Greece. Please do not tell the Greek Cabinet what you are doing and, most of all, don't tell the Minister of War your operations. If you tell one Greek, the Greek character is such that he cannot keep a secret and the enemy will learn about the operation before morning. Please don't even tell me![17]

The general took this advice to heart and worked with and through those members and interests within the Greek government that shared the goals and mission of the JUSMAPG.

Van Fleet believed that the insurgency could not be defeated without the help of a strong national government united behind a "very stable political head" who could "generate a furious love of freedom, a high morale among the civilians on the home front as well as within the armed forces . . . at the front." Eschewing the political government, he turned to the constitutional head of state for this support. The Greek monarchs, King Paul and Queen Frederica, enthusiastically participated in the fight against the insurgents and worked diligently to raise the spirit of the Greek people and the GNA. They founded charitable foundations and camps to support refugees and regularly toured the front and inspected training. Van Fleet embraced the royals and frequently accompanied them on their inspections. This relationship also served as a semi-official channel of communication and cooperation between the supportive elements of the Greek government and Secretary of State Marshall, avoiding the often unhelpful views of the American ambassador.[18]

---

[15] Political Section of the American Embassy Athens, Greece, Political Highlights; Prepared for the Use of Visiting Congressional Delegations, 19 Sep 47, American Embassy Athens.

[16] Iatrides, *MacVeagh Reports, Greece*, pp. 714–20.

[17] Intervs, Col Bruce H. Williams with General James A. Van Fleet, U.S. Army Military History Institute, Senior Officer Debriefing Program, Carlisle Barracks, Pa. (hereafter cited as Van Fleet Intervs), vol. 3, p. 16.

[18] Ibid., p. 21.

This alliance came at a price. The parties of the political left were hostile to the monarchy and saw their relations with General Van Fleet as proof the Americans were "colonizing" Athens. This association also frequently annoyed the U.S. ambassador, Henry F. Grady. Grady had previously served as ambassador to India. He earned a Ph.D. in economics from Columbia and was a committed New Deal Democrat and a friend of President Truman. He disliked Van Fleet and disagreed with his heavy-handed approach to the Communists. Grady urged accommodation and reconciliation over military confrontation.[19]

Gaining the trust and respect of King Paul and Prime Minister Sophoulis allowed Van Fleet to push through an essential reorganization of the Greek Armed Forces (GAF). Together they convinced the National Defense Council (NDC), a cabinet-level body, to grant command authority over the armed forces to General Dimitris Yiatzis, Chief of the Greek General Staff (CGGS). Van Fleet exercised great influence over the CGGS and soon became the dominant voice on the NDC.[20] He had maneuvered into a position of great power, but he realized that he had to act judiciously in exercising his prerogatives. He later asserted,

> I had one great belief—Greece was a sovereign country, and I never imposed myself in violation of their sovereign rights. I think that was something that helped me greatly, whereas, in many of our policies elsewhere we take a superior attitude and assume some of their sovereign rights. This does not go well in getting loyal support and cooperation. So, I always respected their sovereign rights and would never say, "You must do it this way," but rather when necessary, "Here is a wonderful plan that has been worked out by my staff with your staff's concurrence. We ask that you approve it and we'll execute it together."[21]

These relationships, based on shared goals and mutual interests, were strengthened by Van Fleet's willingness to act as an advocate for the Greek cause with the representatives of the U.S. government and the War Department. Many within the U.S. Embassy and the State Department believed that accommodation and reconciliation could best solve the civil war. They pressured the Greek government to adopt many of the reforms demanded by the KKE. Ambassador Grady suggested that aid money spent on security should be redirected into public works and social projects. Washington was impatient and unwilling to increase aid or step up delivery dates. When appropriate, Van Fleet aggressively championed the Greek government's view that security

[19] Ibid., p. 38.

[20] Paul F. Braim, *The Will to Win: The Life of General James A. Van Fleet* (Annapolis, Md.: Naval Institute Press, 2001), p. 167.

[21] Van Fleet Intervs, vol. 3, p. 26.

must come first and that the pace of GNA progress was in many ways linked to the amount and quality of aid received.[22]

While British interests paralleled those of the United States, they diverged over military and political strategy. Although the War Department had directed Van Fleet to cooperate with the British Military Mission to Greece (BMM-G), he did not feel under any obligation to either solicit or follow its advice. By 1948, the British had lost their power to influence events in Greece, and their contributions to the advisory and training effort grew increasingly irrelevant. Nevertheless, the BMM-G and the British ambassador sought to convince Van Fleet to act in accord with British interests. They insisted that U.S. dollars go toward purchasing British weapons and equipment for the GNA. They declared that American strategy was flawed and often "violently" disagreed with Van Fleet's operational advice. He ignored these objections. The American commander was tactful when dealing with the British, proved willing to listen, gratefully accepted whatever practical help the British could offer, and carried on.[23]

Van Fleet functioned as a "general among diplomats" by keeping his focus on the mission while using tact and candor to bring along a collection of willing allies. On this success he would observe,

> If I could offer a bit of advice to an officer of the United States Army who has to work with a foreign leader, either military or political, it would be to avoid condescension as you would the plague. Define the common interest, and let the other party realize for himself how it would be to his advantage to work with you.[24]

Setting the Standard and Demanding Accountability

First, Van Fleet assessed for himself the fighting quality of the GNA. He traveled throughout the country observing operations, talking with commanders and staff officers, and inspecting soldiers and equipment. What he found was not encouraging. The soldiers of the GNA wanted to defeat the Communists, but the army was a poorly trained, tired, and aging force with obsolete equipment and no uniform fighting doctrine. He determined that the army would have to be rebuilt.

It would take time to develop an effective training plan, order and field new weapons, process and train young soldiers, and place them into the depleted ranks of the GNA. More time would be required before these improvements could make a visible impact in the field. Nevertheless, the War Department and the U.S. Army chief of staff demanded to see tangible progress throughout 1948–1949. In particular, U.S. Army Deputy Chief of Staff General J.

---

[22] Briam, *The Will to Win*, p. 172.
[23] Van Fleet Intervs, vol. 3, p. 50; Briam, *The Will to Win*, p. 169.
[24] Briam, *The Will to Win*, p. 166.

Lawton Collins peppered Van Fleet with inquiries regarding the state of training and equipment while simultaneously urging that the Greeks take immediate offensive action against the DAG.[25] The progress report released by the American Mission for Aid to Greece (AMAG) in March 1948 heavily criticized the GNA and caused a stir in the American media. Van Fleet was equally frustrated. He knew that the GNA could not take decisive action until it completed his retraining program.

The insurgents weren't going to grant the Greek government a truce while it reconstructed the national army. The disappointing offensives of 1948, beginning with Operation CROWN in June and Operation VITSI in August to October, served both to illustrate the glaring flaws of the GNA and cement a critical role for the American advisers.[26]

Van Fleet pushed adviser teams down to the division level, and they remained with their assigned Greek units at all times, sharing their hazards and hardships. Each Greek division commander and primary staff officer had an American counterpart. Each adviser team submitted required weekly and monthly status reports detailing the training activities and combat operations undertaken by the assigned unit. These reports allowed JUSMAPG to stay abreast of all developments within the GNA and maintain an accurate tactical and operational assessment of the conflict. With this information and the firsthand knowledge gained from numerous trips around the country, Van Fleet knew more about GNA conditions and operations than the Greek chief of staff. He used this to his advantage when he shared his critical assessments with his counterparts in the Greek Army and the government.[27]

Van Fleet held his advisers responsible for the poor performance of their counterparts in spite of the acknowledged weakness of GNA leadership. Officers and men of the JUSMAPG did not receive this criticism well. They tactfully reminded the general that they did not have command authority over these units, and that in many cases their counterparts had more combat experience. Col. Everett D. Peddicord, adviser to A Corps, said it best: "When the corps commander asked for advice I gave it. Otherwise, I kept quiet. That Greek general knew 10-times more about mountain fighting than I could ever dream of knowing."[28] Most of these men supposed that their job was to provide advice and ensure that American materiel arrived in sufficient quantity and was properly utilized. Van Fleet did not share this view. He expected his advisers to do anything and everything, short of taking part in actual combat, to help their units succeed. As a result, American advisers began skirting dangerously close to full combatant status. In a few instances, they actually took temporary command of Greek units during moments of great crisis. This clearly violated

---

[25] Ibid., p. 172.

[26] JUSMAPG History, pp. 15–16.

[27] Briam, *The Will to Win*, pp. 193–97.

[28] David Colley, "Hot Spots in the Cold War; American Advisors in Greece 1947–49," *VFW Magazine* (May 1997): 36–38.

orders, but Van Fleet either turned a blind eye or imposed token wrist slaps on officers caught making this mistake on a slow news day. Van Fleet expected his advisers to take a proprietary interest in the success or failure of their counterparts, and his men got the message.

The general made accountability an important component of his model of army building. He communicated American displeasure directly and forcefully to the Greek General Staff. Van Fleet gave voice to the frustration he felt over the disappointing results of Operation VITSI through a letter prepared for him by JUSMAPG Deputy Commander Maj. Gen. Reuben Jenkins:

> In all three recent large-scale [enemy] attacks, information on the strength, location, composition and movements of the guerrilla forces was well known by the commanders in the field; yet no offensive action was taken to destroy or seize the initiative from these bandit groups before they struck. . . .
>
> The mission of the Greek Army, as stressed many times in GGS orders, is to pursue and destroy the bandits in Greece and reestablish internal security. The apparent unwillingness of many commanders to accept and understand this basic mission and its clear implications is responsible for the failure of the Greek Army to take aggressive offensive action against bandit groups, pursue them, and destroy them wherever they may be.
>
> Practically all commanders invariably plead unfavorable weather, forbidding terrain, inadequate strength and serious dangers to vital areas, as a defense for their inertia and failures. Yet, the inferior bandits continue to operate in the same weather, and over the same terrain as must the Greek Army, with high success, in complete defiance of, and with bold indifference to, a greatly superior, better fed, better clothed, better organized and better equipped Greek Army. This is a definite reflection upon the quality of Greek Army leadership in the field.[29]

The inadequate performance of the GNA did not go unnoticed by American civil leadership in Athens and Washington. Ambassador Grady speculated that the GNA deliberately held back on its efforts to defeat the insurgency in order to coax more aid dollars from the United States.[30] President Truman admitted to Congress that the war in Greece had not advanced beyond the point of military stalemate, "despite the delivery of $170,000,000 of United States arms supplies and the advice of a United States military mission."[31]

Van Fleet made it clear to the Greeks that U.S. aid was conditional and ephemeral. If the GNA could not demonstrate a willingness to take the fight to the Communists, that aid would end. They were all accountable to the American taxpayers and their elected representatives. He urged the Greek General Staff to relieve senior commanders who had underperformed during the

---

[29] Briam, *The Will to Win*, p. 205.

[30] "Scant Results from U.S. Aid to the Greeks," *U.S. News & World Report*, 29 Oct 48, p. 24.

[31] "Truman Reports Greek Stalemate," *New York Times*, 7 Dec 48.

1948 offensives. Action came fast. Van Fleet had several senior commanders relieved and persuaded General Yatsis, chief of staff of the Greek Armed Forces, to retire.[32]

On 21 January 1949, the Greek government appointed General Alexander Papagos commander-in-chief of the GNA. Papagos was a national hero and veteran of the 1940–1941 Albanian Campaign against Fascist Italy. An aggressive and competent commander, he was the preferred choice of both the king of Greece and Van Fleet. This change at the top had a significant, positive impact on the fighting abilities of the GNA. As the official history of the advisory group states,

> This was a very important change in the Greek High Command, for it influenced decisively all future GAF activities. By enforcing much needed discipline in the senior ranks of the Army, by eliminating incompetent and insubordinate officers, by stressing mobility and aggressiveness, and by insisting that units constitute reserves and wholeheartedly accept the JUSMAPG-sponsored training program, General Papagos initiated accomplishment of many of the aims which JUSMAPG had long advocated.[33]

While disagreements would remain regarding the size of the GNA and the amount of funding required, the Greeks had come to realize that the Americans had limited patience and would make continuing aid contingent on positive results.

Rebuilding the Greek National Army

Van Fleet was determined to rebuild, retrain, and reequip the Greek National Army according to the unique requirements of the conflict.

The first step in this process was to refresh the ranks of the GNA with younger, loyal, and more capable recruits. Naturally, some of the soldiers inducted into the army sympathized with the Communist cause. The Greeks, of course, knew this, and they implemented a program to weed out politically undesirable recruits at the induction stations. Conscripts determined to be politically unreliable were sent to training and labor battalions on the Island of Makronisos, where they underwent political indoctrination and received basic military training. Opinions are mixed on whether this measure was successful. There are a few accounts of the king and queen visiting these camps and being carried around the perimeter in triumph on the shoulders of the inmates. Others maintain that these were little more than internment camps. It is never pleasant to be deprived of your liberty or forcibly separated from loved ones, but the men appear to have been decently treated. Van Fleet believed these camps were required to break up Communist intelligence cells and dry up the pool

[32] Briam, *The Will to Win*, pp. 188–89.
[33] JUSMAPG History, p. 23.

of potential DAG recruits. To boost soldier morale, Van Fleet encouraged the GNA to establish a postal service so the men could keep in contact with their families. This news from home had the added benefit of countering Communist propaganda that claimed the civilian population was starving and suffering great privations to support a war of American imperialism. Older soldiers were retired from active service or placed in reserve formations and replaced with fit, fresh young men. Perhaps most important, Van Fleet pushed the Greek government to build a national army composed of citizen-soldiers committed to the cause of a free and democratic Greece. He also made clear the need to fill the ranks with intelligent, educated soldiers capable of using and maintaining the new American weapons and equipment pouring into the country. Three classes of fifteen thousand soldiers each were cycled through an intense, six-week basic training program and posted to the ranks of the field divisions in the winter of 1948.[34]

JUSMAPG developed an intense, battle-focused training program that emphasized fundamental soldier skills and culminated in company- and battalion-level exercises. Unit training emphasized close cooperation among infantry, field artillery, and close air support. Much of the training was conducted with live ammunition. U.S. and Greek instructors taught GNA soldiers to use indirect fire to suppress bunkers and defensive positions and to advance and attack directly behind this fire. Soldiers maneuvered on training ranges where they were required to engage targets to their flanks and rear with direct fire. Van Fleet knew that the Greek government was willing to accept the inevitable accidents and casualties that resulted from this type of aggressive training. He later recounted, "Their training was even better than we dare give American troops. We could work them from dawn to sunset six days a week."[35] The program emphasized night movement and aggressive patrolling techniques at the platoon and company level. Newly raised demonstration platoons provided examples of these techniques to units in the field. The training placed an emphasis on aggressive action and the need to demand results up and down the chain of command.[36] In his written report to General Papagos, Van Fleet wrote,

> Every day of free time must be devoted to training. Commanders must not be satisfied with mediocre results; they must demand and receive the best. In combat they must be prepared to pursue relentlessly day and night, regardless of local boundaries. They must have, and must instill in their commands, the will and burning desire to close with the enemy at all times.[37]

[34] Van Fleet Intervs, vol. 3, p. 20.

[35] James A. Van Fleet, "25 Divisions for the Cost of One," *Reader's Digest*, 1954, p. 3.

[36] Briam, *The Will to Win*, pp. 183–85.

[37] Ibid., p. 191.

The primary problem remained an unwillingness or inability to close with and destroy the insurgents. He sought to overcome this deficiency by instilling aggressive leadership through hard training and by the judicious purchase of critical weapons and equipment. Van Fleet was confident that the combat performance of the army would improve as a result of these measures, but he maintained realistic expectations. This focus on fundamentals brought him into conflict with members of the joint group who pressed for the adoption of more modern and expensive equipment and techniques. The best example of this was the proposal by JUSMAPG Air Force Chief General Metheny to purchase large numbers of helicopters and form air mobile units capable of landing in the remote and rugged mountain areas where the guerrillas built their base camps. Van Fleet thought the scheme absurd and stated,

> General, it isn't getting the army up there; they can march up there just as quickly by starting near the base of the nearest location and climbing the mountain on foot with mule supply. The job is to get them to move out promptly after they get there. We would have the same problem whether we were carried there on foot or by helicopter.[38]

This was a source of great frustration to Van Fleet. He preferred that his senior airman focus on improving cooperation between close air support assets and ground units, but Metheny continued to push his air assault scheme. In the end, Van Fleet fired him and bought pack animals instead. JUSMAPG requested P–51 fighter-bombers to provide close air support for the resurgent Greek Army, and Van Fleet was furious when these planes were sent to Turkey instead. The Greeks got SB2C Helldivers. Van Fleet grumbled, but he made do with the means provided.[39]

He expended the limited military credits available on weapons that would help the infantry overcome the strong pillboxes and fieldworks that had consistently stymied the squads and platoons of the Greek Army. The obsolete field guns of the GNA could not adequately suppress the reverse slope defense favored by the DAG. The Greeks needed pack howitzers, rocket launchers, and recoilless rifles. Van Fleet pushed hard to get them from a stingy and skeptical War Department.[40]

The British had provided the GNA with sound counterinsurgency doctrine, but the officers of the Greek Army did not internalize or follow it. The preferred technique was to approach the insurgent positions, move into blocking positions, and then attempt to annihilate them with indirect fire. Local commanders often demonstrated the obnoxious and often disruptive habits of quarreling with superiors or refusing to obey orders they disagreed

---

[38] Van Fleet Intervs, vol. 3, p. 44.

[39] Ibid., pp. 45–46.

[40] Briam, *The Will to Win*, pp. 209–10.

with. Van Fleet attributed this to the British military culture that dominated the Greek officer corps:

> They had a British system of giving orders. . . . For example, should a (unit) commander . . . not like the order received, he does not have to obey it promptly. He can appeal to the next higher command. If the higher commander approves the original order then he has to carry it out. Or, should his own commander give it to him a second time in writing, he is obliged to carry it out. That slows down the execution of orders—and in chasing guerrillas or trying to surround them in a mountainous area and move in. You can't have such a system of command. You drag your feet and the enemy gets away.[41]

This and other factors combined to make the GNA slow moving and tactically timid. JUSMAPG advisers at every level sought to overcome this lethargy by preaching the "5 F's: Find—Fend—Fix—Fight—Finish."[42] Van Fleet insisted that the army move at night and during periods of limited visibility. The Americans advised the Greeks to move their best troops, under cover of darkness, into blocking positions astride the insurgents' route of escape and launch dawn attacks with units of lesser quality to drive the insurgents from their positions and into prepared ambushes. Van Fleet urged the Greek commanders to press the attack in the winter and through inclement weather and reminded them that as the better equipped, fed, and trained force, they had an advantage over the semi-starved and shivering insurgents. The Americans stressed junior leader initiative, aggressive pursuit, and combined arms to their counterparts. Van Fleet's men did not convert everyone, but they did manage to convince the Greeks to take the fight to their enemy in a way they had not done before. By summer 1949, JUSMAPG reported that

> The Greek armed forces retained the initiative over widespread areas. DAG tactics when confronted with any force were generally those of dispersal, evasion, defense and withdrawal. GNA pursuit in offensive areas kept these groups unsettled and moving.[43]

The rebuilt Greek Army was far from perfect, but it was capable enough to take advantage of the mistakes made by the DAG and defeat it in open battle in the remote mountains of northern Greece.

Making It "Their War"

The United States government believed it had important national security interests in Greece, but few regarded the stakes as high enough to warrant the

---

[41] Van Fleet Intervs, vol. 3, p. 12.
[42] Van Fleet, "How We Won in Greece."
[43] JUSMAPG History, p. 24.

introduction of American combat troops. JUSMAPG directed that advisers would not carry weapons or engage in combat.[44] They were not to assume command of Greek troops or engage in political activities. As far as Washington was concerned, this was a civil war, an internal conflict that would be won or lost by the Sophoulis regime.[45]

The Greeks saw things differently. Most of their military and political leaders thought of the war as part of the larger conflict between the United States and the Soviet Union, and American threats and reluctance to provide additional aid often puzzled and outraged them. Many Greek citizens believed they were little more than American pawns being ground up in a conflict between the two superpowers.[46]

Van Fleet thought that both sides were correct. He agreed with his Greek counterparts that they faced an attempt by the Soviets to conquer Greece through the instigation and support of an insurgency against the legitimate government. He also held that it was the responsibility of the Greeks to defeat this threat to their liberty.[47] The DAG certainly received supplies and sanctuary from sympathetic allies in Yugoslavia and elsewhere, but the fact remained that the insurgents were their fellow countrymen and women. On the American role, Van Fleet explained,

> We are here to administer material assistance in an advisory capacity only. We do not command or fill any executive positions. The decisions and the performance of the [Greek Armed Forces] are entirely Greek matters.[48]

It was imperative for the Greek government to break the Communists' hold over the rural population and eradicate, or at least cripple, their urban espionage and support cells. The operational success of the GNA depended on it. Van Fleet knew that this could only be accomplished by the Greeks themselves. The gendarmerie made use of mass arrests and preventive detentions to weed out DAG sympathizers and collaborators. Greek internal security forces, accountable to the population and operating under Greek civil law, interrogated and punished leaders. These measures were certainly harsh, but Van Fleet believed they were justified under the circumstances. This is not to say that he countenanced violations of the Geneva Convention or approved of the mistreatment of prisoners, but he saw a distinction between an insurgent under arms in the field and a terrorist informant. These tactics did not pass unnoticed by the American press or the U.S. Embassy, and displeasure with these actions often caused friction between Van Fleet and the ambassador.

---

[44] JUSMAPG, Duties of JUSMAPG Personnel in the Field, 29 Apr 48, Athens.

[45] JUSMAPG, Memo on the Organization and Functions of JUSMAPG, 22 Apr 48, Athens.

[46] "War Risks in Greek-Aid Plans," *U.S. News & World Report*, 5 Mar 48, pp. 30–31.

[47] Van Fleet, "How We Won in Greece."

[48] "Van Fleet Defines Threat to Greece," *New York Times*, 29 Nov 48.

Nevertheless, the JUSMAPG commander steadfastly supported these measures because he saw them as the most timely and effective way to neutralize the Communist intelligence network.[49]

Van Fleet's plan to rebuild the citizen militias also generated controversy. At the start of the conflict, the Greek government created ad hoc groups of armed citizens to provide local security in the small villages that dotted the mountains and valleys of the country. These formations were equipped with captured German and Italian small arms and controlled by local officials. These militias had failed to prevent insurgent raids, and many of the weapons provided ended up in the hands of the Communists. The inability of the local population to defend their own homes and villages meant that the GNA had to provide garrisons to augment the overstretched gendarmerie. Frightened and largely defenseless civilians fled the countryside, flocking to the larger towns and refugee camps. This damaged the agrarian economy and the social fabric of the country.

However, Van Fleet believed that the loyal citizens of Greece would take up arms to protect their homes and property. He ordered that these militias be rebuilt as Home Guard detachments. The Greek government equipped the militia with British small arms, made surplus and obsolete by the arrival of new American weapons, and placed them under the strict control and supervision of the GNA. According to the official history of the JUSMAPG, this measure "contributed materially to the economic recovery of the country as large numbers of refugees left the government dole and entered productive fields." This step also brought a substantial part of the rural population over to the side of the government, making it more difficult for the insurgents to intimidate and exploit them.[50]

Unlike the British, who had been involved in Greek affairs since the early nineteenth century, the United States had few interests in the eastern Mediterranean prior to World War II. The JUSMAPG lacked the detailed knowledge of the region and the assets required to collect the vital intelligence data necessary to defeat the insurgents. The Americans relied on the Greeks to do the majority of intelligence collection. Van Fleet explained,

> You cannot do anything without intelligence. You must know where the enemy is; his possible intentions. You have to invade his unit with a spy or spies and have that information brought back. . . . The Greeks are very good (intelligence) people. They had many agents all over the Balkans. . . . They had good information as to what was going on in these countries in support of the communist guerrillas; such as hospitals, training grounds, supply elements, rest areas and the like. . . . They had their own people scattered

[49] Van Fleet Intervs, vol. 3, pp. 49–52.
[50] JUSMAPG History, pp. 11–13.

> around in the towns, cities and villages of Greece. . . . The Greeks have a way of getting information.[51]

The JUSMAPG staff provided assistance in the analysis and effective use of this intelligence, but Van Fleet left the difficult and potentially compromising job of collection to the Greeks themselves.

Defeating the Communist insurgency was a Greek responsibility, and Van Fleet never wasted an opportunity to remind them of this fact. He appealed to their sense of pride and national honor. While inspecting Greek soldiers in the field, he would stand before the ranks, turn to his entourage, and say,

> I wish that you would look closely at this Greek soldier, for he is a member of one of the bravest and most distinguished armies of the entire Free World. He is a patriot fighting for the freedom of his country, and I am proud to be associated with him in this cause. . . . I am proud to be helping an army that is fighting the communists on its own, helped only by American military equipment.[52]

Conclusion

The American mission to Greece did not achieve all of its objectives. However, it did achieve its primary goal of assisting Greece to "preserve its independence."[53] Much of the credit for this must go to James Van Fleet and the men of the JUSMAPG. In Van Fleet's own words,

> Our initial $300,000,000 investment, made in 1947, has paid off magnificently. . . . Greece was saved for little more than the cost of a single American division. Yet not one American infantryman had ever fired a shot on Greek soil.[54]

He believed that the American approach in Greece was certain to work elsewhere around the world and urged that it become a cornerstone of America's strategy to defeat the Soviet Union. He advocated that the Army construct "packages" of equipment and advisers specially configured for individual nations under threat and "dot the iron curtain" with them, providing threatened allies with the means to fight for their own freedom:

> We cannot furnish the world with American ground armies. If we try, it will drain us of manpower and throw us into bankruptcy without ever firing a shot, which is what the Soviets want. . . . We should stand ready, when the communists poke into a soft spot, to provide these native divisions, defending

---

[51] Van Fleet Intervs, vol. 3, p. 21.
[52] Briam, *The Will to Win*, p. 191.
[53] AMAG Survey, p. 2.
[54] Van Fleet, "25 Divisions for the Cost of One," p. 10.

> their native land, with the munitions of war they will need but cannot make for themselves.[55]

Van Fleet firmly believed in the Truman Doctrine. The experience in Greece, combined with his successful effort in rebuilding the army of the Republic of Korea during the Korean War, convinced him that helping to raise, equip, and train the forces of friendly governments under threat of invasion or subversion should become a core competency of the U.S. Army. Perhaps it is time for us to reconsider this assertion.

---

[55] Ibid.

Chasing a Chameleon: The U.S. Army Counterinsurgency Experience in Korea, 1945–1952

Mark J. Reardon

In his widely read book, *On Strategy: A Critical Analysis of the Vietnam War*, Col. Harry G. Summers Jr. argues, "We failed in Vietnam because we attempted to do too much. Instead of concentrating our efforts on repelling external aggression as we had done in Korea we also took upon ourselves the task of nation building."[1] Not everyone, however, supported that particular postmortem assessment. Air Force Col. Kenneth J. Alnwick believed Vietnam had been "essentially unwinnable" because "the Saigon government, a creature of the United States, consistently demonstrated its inability to resolve its internal contradictions, to govern South Vietnam, and simultaneously to prosecute a protracted war against a dedicated, determined, enemy."[2] Indeed, Colonel Alnwick suggests that American preoccupation with conventional combat, rather than strengthening the Saigon government, contributed to eventual defeat.

Summers felt extremely comfortable with citing Korea as an example for fellow Army officers to heed. After all, he served in the 6th Infantry Division in Korea during 1947–1948 and later fought there with the 24th Infantry Division. Perhaps because he lacked comparable Korean experience, Alnwick did not contest Summers' assertion that American ground forces focused on conventional combat during that conflict. Additional research by Alnwick, however, would have disclosed that the U.S. Army's role in Korea was not confined to conventional combat operations.

American involvement began almost five years before the Communist invasion orchestrated by Kim Il Sung, when Emperor Hirohito's acceptance of Allied demands for unconditional surrender on 15 August 1945 stunned Japanese military personnel and civil servants in Korea. The quarter-million-strong Japanese garrison on the peninsula had not been defeated in battle. Indeed, few, if any, had ever seen an American. With the exception of several small-scale carrier raids, the Pacific War bypassed Korea.

---

[1] Harry G. Summers Jr., *On Strategy: A Critical Analysis of the Vietnam War* (Novato, Calif.: Presidio Press, 1982), p. 171.

[2] Kenneth J. Alnwick, "Strategic Choice, National Will, and the Vietnam Experience," *Air University Review* 34, no. 2 (January–February 1983): 136.

The Korean people, who had endured decades of harsh rule, were overjoyed at the news. The Japanese initially "established their hegemony over Korea via the Treaty of Shimonoseki and dictated to the Korean government a wide-ranging series of measures to prevent further domestic disturbances" following the Sino-Japanese War in 1894.[3] The Russo-Japanese conflict of 1904–1905, which ended in Russia's defeat, led to Japan's complete annexation of Korea on 22 August 1910.

Although grateful for American sacrifices that restored their freedom in 1945, virtually every Korean desired immediate and full independence. Within hours of the emperor's announcement, they began holding mass meetings and parades. Communist-backed political groups, having received significant material and financial assistance from the Soviets over the past two decades, were particularly in evidence.[4] Celebration gave way to rioting and looting, which triggered a ruthless response by the Japanese.

The American occupation force, consisting of XXIV Army Corps led by Lt. Gen. John R. Hodge, did not land at Inch'on until 8 September.[5] With most specialized civil affairs units earmarked for Japan, Hodge had converted the Tenth Army Antiaircraft Artillery headquarters into a military government staff.[6] The commanding general of the 7th Infantry Division, Maj. Gen. Archibald Arnold, served as the first American military governor of Korea in addition to his regular duties.[7]

The Allied powers were reluctant to immediately address the question of Korean independence. A wartime agreement bisecting the peninsula at the 38th Parallel played no small part in complicating matters. The Russians occupied the peninsula north of the dividing line, while the Americans occupied the south. Neither superpower wanted an independent Korea to fall within the sphere of influence of its former ally turned postwar ideological adversary. While the Russians and Americans jockeyed for strategic advantage, a similar contest began in southern Korea between the leftist People's Republic Party and the right-wing Democratic Party.

With Hodge's troops spread thinly across the southern half of the peninsula, establishment of an indigenous security force became one of the top American priorities. In a 23 September 1945 report, the XXIV Corps provost marshal, Brig. Gen. Lawrence E. Schick, proposed the creation of a Korean Political

---

[3] HQ, Department of the Army Pamphlet 550–41, *South Korea: A Country Study* (Washington, D.C.: Government Printing Office, 1992), p. 14.

[4] Daniel R. Shea, "An Analysis of the United States Leadership's Effective Employment of the Republic of Korea Armed Forces, July 1950–June 1951" (Master's thesis, U.S. Army Command and General Staff College, Fort Leavenworth, Kans., 2001), p. 17.

[5] History of the United States Army Forces in Korea, vol. 1, ch. 1, p. 6, Historical Reference Collection, U.S. Army Center of Military History (CMH), hereafter cited as Army Forces in Korea.

[6] Ibid., p. 27. The 6th Division replaced the 96th Division on 23 September 1945.

[7] Shea, "An Analysis of the United States Leadership's Effective Employment of the Republic of Korea Armed Forces," p. 20.

Affairs Department (as it was then called) consisting of administrative, finance, criminal investigation, clothing and supply departments, and a training school.[8] The organization was later renamed the Korean National Police.

The military government's efforts to create an indigenous internal security force produced unforeseen results. The Americans initially recruited former Korean members of the Japanese Army and the colonial police.[9] As one American explained, "A Korean peasant police recruit, living in an ox-cart society, knew only Japanese police methods. He was brought up in the belief that there was no government without terror, that a prisoner was guilty until proven innocent, and that torture should precede the examination of witnesses to insure that only the truth would be told."[10] Efforts to teach democratic ideals and individual rights to the National Police were hampered by a chronic lack of American supervisors and instructors.

Projected food shortages forced the military government to impose a highly unpopular rice collection program in December 1945. The Americans made the newly created National Police responsible for collecting rice from farmers and delivering it to central storage sites. The collection effort met with widespread protests, which produced violent responses by Korean police. The unpopularity of the police increased when the American Military Government, following orders issued by General Douglas MacArthur's headquarters in Tokyo, ordered them to recover Japanese property confiscated by Korean civilians.

Ongoing problems with the National Police, compounded by growing American suspicion of Soviet aims, prompted General Hodge to reorganize and expand South Korean security forces. Armed with broad guidance from Hodge, a board of American officers chaired by General Schick met on 10 November to devise a draft plan. Schick subsequently presented Hodge with a proposal for an Army and Air Force of 45,000, a Coast Guard of 5,000, and a National Police Force of 25,000.[11] Accepting the majority of the board's recommendations, Hodge appointed Schick as the first director of national defense for Korea on 13 November.

American plans for creating the South Korean Army gained momentum as widespread rioting broke out in Namwon on 15 November. The protests began when police ordered locals to return stolen Japanese property. The citizens refused, and the police arrested several of them. The arrests ignited a wave of violence directed at the police, who called for American assistance. When U.S. troops arrived, they were confronted by an unruly crowd of one thousand rioters. The Americans fired a volley into the air before dispersing

[8] Army Forces in Korea, vol. 3, pt. 1D, p. 8.

[9] Ibid., p. 9.

[10] Ibid., p. 41.

[11] Richard P. Weinert, The U.S. Army and Military Assistance in Korea Since 1951, OCMH Manuscript 104-M, n.d., p. 7, Historical Reference Collection, CMH.

the mob with leveled bayonets. The ensuing melee killed or injured a number of Koreans, including one policeman stabbed to death.[12]

The events at Namwon reflected a growing dissatisfaction with the National Police and undiminished Korean desire for immediate and complete self-rule. Korean nationalists expressed that belief so frequently and vehemently that General Hodge reported to MacArthur on 16 December,

> Under present conditions with no corrective action forthcoming, I would go so far to recommend we give serious consideration to an agreement with Russia that both U.S. and Russia withdraw forces from Korea simultaneously and leave Korea to its own devices and an internal upheaval for its self purification.[13]

The first tentative steps toward Korean self-rule began when the United States and the Soviet Union convened a joint commission in December 1945. Talks on the peninsula's political future, which did not begin until March 1946, ended two months later. The negotiations broke down when the Russians refused to withdraw their proposal to limit future membership in a Korean provisional government to political parties that accepted trusteeship prior to full independence. Since every right-wing political party rejected trusteeship, the Russian initiative would have created a Communist-dominated political system.

The stalled talks had a significant impact on Hodge's initial plan to improve internal security in South Korea. Concerned about possible Russian protests, the State Department asked Hodge to postpone creation of a South Korean Army. Rather than suspend his efforts, Hodge restructured the army as a police reserve (National Constabulary) and reduced its size to 25,000.[14] The constabulary's primary mission would be to support local police in a time of disorder or emergency, whenever the military governor or his Korean counterpart called for such aid.[15]

Hodge forwarded the revised plan to Washington, where the Joint Chiefs of Staff approved it in January 1946.[16] Hodge immediately transferred eighteen lieutenants from the military government to serve as the initial contingent of administrators and trainers. The Americans also supplied the constabulary with

---

[12] Bryan Robert Gibby, "Fighting in a Korean War: The American Advisory Missions from 1946–1953" (Ph.D. diss., State University of Ohio; Ann Arbor, Mich.: University Microfilms, 2004), p. 30.

[13] Shea, "An Analysis of the United States Leadership's Effective Employment of the Republic of Korea Armed Forces," p. 21.

[14] Ibid., p. 31.

[15] Army Forces in Korea, vol. 3, pt. 1A, p. 69.

[16] Gibby, "Fighting in a Korean War: The American Advisory Missions from 1946–1953," p. 33.

Japanese arms and equipment. No plans were made to provide it with heavy weapons such as tanks and artillery.[17]

Recruiting the constabulary's leadership constituted the first priority for the Americans. The military government invited representatives of militia groups and selected individuals to a meeting held at the Old Capital Building in Seoul. They were greeted by Col. Reamer W. Argo, who encouraged them to join the constabulary as commissioned officers. Many attendees agreed to join upon learning that the Americans prized past military experience, even under Japanese tutelage, over current political affiliation.

The recruiting of enlisted personnel proved more difficult. American advisers noticed that little popular sentiment existed for an indigenous army.[18] Inadequate incentives had created a perception among Koreans that service in the ranks offered low pay and benefits. By the end of April 1946, the constabulary numbered but two thousand officers and men.[19] However, recruiting improved during the following month when the National Police started rounding up hundreds of Communist activists.[20]

The changing fortunes of enlisted recruiting reflected the growing political extremism in Korea. The members of the Korean National Police, most of whom possessed right-wing sympathies, played a large part in creating this environment. Police officers arrested leftists while openly shielding right-wing groups from retaliation. As political violence increased, the American Military Government noted, "It seemed that any political figure who cut loose from the Extreme Rightist and Extreme Leftist camp and attempted to steer a middle course simply committed political suicide."[21]

The suspension of Soviet-U.S. talks in May 1946, combined with the departure of the 40th Division, persuaded left-wing Korean groups to adopt a strategy calling for "an all-out attempt to discredit the American effort, to sabotage Military Government, and to build up hatred against the Americans."[22] The American Counter Intelligence Corps (CIC) responded by redoubling efforts to penetrate and disrupt Korean extremist groups. A raid in Kongju on 18 July netted a document signed by a provincial committee requesting information "relative to American Forces." Similarly worded instructions were also found in Pusan.[23] On 7 August, CIC personnel raided the home of the chief of political affairs for the South Korean Communist Party. The Americans recovered numerous documents, to include "detailed but not too reliable

[17] Weinert, The U.S. Army and Military Assistance in Korea Since 1951, p. 7.

[18] Gibby, "Fighting in a Korean War: The American Advisory Missions from 1946–1953," p. 35.

[19] Army Forces in Korea, vol. 3, pt. 1A, p. 69.

[20] Gibby, "Fighting in a Korean War: The American Advisory Missions from 1946–1953," p. 35; Army Forces in Korea, vol. 2, p. 344.

[21] Army Forces in Korea, vol. 2, p. 329.

[22] Ibid., p. 330.

[23] Ibid., p. 339.

information on United States Army installations and units, detailed information on the Korean Constabulary and Coast Guard, Military Government, and the police. Translations of these documents revealed a most successful infiltration into the Constabulary and Coast Guard."[24]

The CIC raids failed to uncover information on a nationwide railroad strike on 23 September 1946 that heralded the Communist-inspired "Autumn Harvest Rebellion." The striking railway workers were joined a few days later by printers, Seoul electric company employees, and postal workers in Seoul, Taegu, and Kongju. The labor stoppages preceded in an orderly and nonviolent manner for six days before the death of a rioter in Taegu altered the mood of the demonstrators.

On 2 October, a large mob stormed the Taegu main police station. The police ran for their lives rather than mount an effective defense. Several were caught, tortured, and killed. When American civil affairs personnel requested assistance from Col. Russell J. Potts of the 1st Infantry Regiment, they were dismayed to learn that he refused to send any U.S. troops into Taegu unless he received full authority to declare martial law.[25]

American Military Government officials in Taegu received a second surprise when Korean constabulary units remained in their barracks rather than halt the rioting.[26] The constabulary officers were worried that their enlisted ranks, which contained many leftist sympathizers, would refuse to obey if told to fire upon the rioters. The belated arrival of U.S. troops and armored vehicles forced the mob to relocate to the city's outskirts, where rioters continued to wreak havoc and set fire to buildings. When the rioters refused to disperse peacefully, the Americans opened fire, killing or wounding several Koreans.[27] Although order was restored in Taegu, an American report noted that the uprising proved "sufficiently successful to serve as a model for future serious rioting and to spur on the Communists in further fanning the flames of disorder."[28]

Violent demonstrations were not confined to Taegu. At Waegwan, two thousand rioters overran the main police station on 3 October. After killing the chief of police, the mob seized all the weapons and ammunition they could find. One group loosed a few ineffective shots at a U.S. patrol responding to the incident.[29] Four days later, American soldiers opened fire on mobs in Chinju

---

[24] Ibid., pp. 341–42.

[25] "The Quasi-Revolt of October 1946," Army Forces in Korea, vol. 3, pt. 1C, p. 34. Potts received an official admonishment for his failure to act promptly.

[26] Allan R. Millett, *The War for Korea, 1945–1950: A House Burning* (Lawrence: University Press of Kansas, 2006), p. 85.

[27] Political Adviser in Korea (Langdon) to Secretary of State, 1 Nov 46, in *Foreign Relations of the United States (FRUS), 1946* (Washington, D.C.: Government Printing Office, 1971), 8:754.

[28] "The Quasi-Revolt of October 1946," p. 1.

[29] Ibid., p. 5.

and Masan, killing twelve Koreans and wounding more than one hundred.[30] On 17 and 19 October, Korean police requested American assistance at Myonchon and Yeson, respectively.

On 22 October, the American Military Government dispatched tank patrols into Seoul to prevent rioting.[31] Nine days later, elements of the 20th U.S. Infantry Regiment and Korean police fired on a 5,000-strong mob converging on Nanju, killing ten dissidents and injuring dozens.[32] More violence occurred in early November, to include an attack near Kwangju against a U.S. military government convoy that resulted in Korean deaths and injuries to two Americans and thirty-three Koreans.[33]

The Autumn Harvest Rebellion quashed U.S. desires to maintain an occupation force in Korea. Shortly afterward, the State Department circulated a memorandum discussing possible timelines and conditions for an American withdrawal. The assistant secretary of state for occupied areas told the director of far eastern affairs,

> I agree with you that it would be far better to have a unified central Government before we withdraw. . . . My judgment is that it isn't the existence of a central government that is of such importance to us, but rather the existence of a reliable, strong, nationally-recognized central government. If we wait for that to occur, our troops may be there for ten years. . . . I think if the Russians would come forward tomorrow with a proposition for both of us to pull our troops out of Korea, we would decide—and very properly in my opinion, to haul our freight.[34]

Social unrest remained high in Korea as 1946 drew to a close due to widespread unemployment (only half of the labor force of ten million had jobs), food shortages, and frequent electrical blackouts. Poor living conditions had produced, "disillusioned and disconcerted people [who] paid keen attention to political leaders of various persuasions who offered new ways of solving Korean problems."[35] The South Korean Communist Party tapped into the well of popular discontent to stage a series of violent demonstrations against the

---

[30] Millett, *The War for Korea,* p. 85, and "The Quasi-Revolt of October 1946," p. 6. An American report placed the death toll in Chinju alone at four killed and "many more" wounded.

[31] Political Adviser in Korea (Langdon) to Secretary of State, 14 Nov 46, in *FRUS, 1946*, 8:766.

[32] "The Quasi-Revolt of October 1946," p. 13.

[33] Ibid., p. 14.

[34] Memo, Assistant Secretary of State for Occupied Areas for Director of the Office of Far Eastern Affairs, 8 Nov 46, in *FRUS, 1946*, 8:764–65.

[35] *South Korea: A Country Study*, p. 30.

American presence in February 1947.[36] On this occasion, U.S. combat units stayed in their barracks.[37]

Spurred by a growing desire to disentangle itself from an unpopular international commitment, the United States brought Korean independence before the United Nations (UN) General Assembly in September 1947. Within a month, the UN initiated plans to create a Korean civilian government prior to the departure of the American occupation force. Realizing they lacked the votes to prevail in a nationwide contest, North Korean Communists announced their intention to boycott the UN-sponsored election completely. South of the 38th Parallel, political violence increased as right-wing factions sought to enhance their influence prior to national elections scheduled for May 1948. The South Korean Communists focused their efforts on undermining the legitimacy of the upcoming election.

The Cheju-do Rebellion of April 1948 preceded the Korean national election by one month. The American Military Government decided not to employ U.S. combat troops to restore order. Taking a cue from the "hands-off" attitude of their superiors, the military government officials on Cheju-do contributed to a widening of the crisis when, "the Chief Civil Affairs Officer of the 59th Military Government Company failed to take prompt and determined action to control the island police, and [to] effectively employ the police reserves as they arrived on the island."[38] Poorly coordinated sweeps into the island's mountainous interior resulted only in numerous police casualties. Angered by their losses, the police vented their frustration on innocent villagers, which only served to persuade undecided individuals to join the insurgents. The 9th Constabulary Regiment on Cheju-do remained in its barracks. Its commander radioed Seoul to request instructions and to report that his soldiers were afraid of the rebels.[39]

Determined to ensure the islanders participated in the national elections, the American Military Government deployed the 11th Constabulary Regiment to Cheju-do. Capt. Jimmie Leach, who served as senior adviser to the unit, recalled, "I had six [other Americans] with me; I could call on two small L–4 single engine, fabric covered scout planes, and I had a navy too; two old, wooden-bottomed minesweepers converted to coastal cutters and manned by Korean crews."[40]

The 11th Regiment began systematically hunting down insurgent bands. The National Police, heartened by the presence of an aggressive constabulary

---

[36] Korea in War, 500th Military Intelligence Group, translation requested by OCMH, Document 80545, n.d., p. A-166, Historical Reference Collection, CMH.

[37] Harry G. Summers Jr., The Korean War: A Fresh Perspective, www.rt66.com/~korteng/SmallArms/24thID.htm, Historians files, CMH.

[38] Army Forces in Korea, vol. 3, pt. 1, p. 20.

[39] "Back in the Day—Col. Jimmie Leach, a Former U.S. Army Officer, Recalls the Cheju-do Insurrection in 1948," by Col. Jimmie Leach as told to Matt Hermes, *Beaufort Gazette*, 10 Jan 2006, www.beaufortgazette.com, Historians files, CMH.

[40] Ibid.

unit, stepped up its efforts to destroy the rebels. The rebellion also spurred American efforts to create an all-sources intelligence cell to "collect, coordinate, analyze, and disseminate all insurgency related information from police, military, and civilian sources."[41] The first cell proved so effective that the Korean Military Assistance Group (KMAG) directed its advisers to create similar organizations in each Korean division.

Although it took almost a year to finally defeat the Cheju-do insurgents, the uprising did not derail the May 1948 elections. Special precautions, to include the arrest of 993 leftist agitators by American CIC agents and Korean police, served to reduce violence directed at voters and polling places.[42] Not surprisingly, the hard-line rightists won an overwhelming victory. Three weeks later, the constitutional assembly appointed Syngman Rhee as its chairman. Rhee's personal agenda consisted of solidifying his power base rather than instituting a truly democratic political system. Rhee cooperated with the American Military Government only to the extent necessary to further his own ambitions.

Events in Korea moved rapidly in the wake of the elections. On 15 August 1948, the Republic of Korea (ROK) was formally established. That same day, the constabulary was renamed the Republic of Korea Army. Three weeks later, Kim Il Sung assumed the reins of power in North Korea, claiming "authority over the entire country by virtue of elections conducted in the north and the underground elections allegedly held in the south."[43]

Kim Il Sung's ascension to power coincided with plans to increase the South Korean Army to 50,000 officers and men. While the number of American police advisers shrank to four, the U.S. contingent of Korean Army advisers increased from 100 to 248.[44] Brig. Gen. William L. Roberts, who had replaced General Schick, established the Provisional Military Assistance Group, Korea, with a table of organization that would provide the South Koreans with American counterparts down to battalion level. The departure of the last American combat troops, scheduled for June 1949, also served to accelerate U.S. plans to reconfigure the South Korean Army into a division-based organization.

The South Korean Army faced a number of challenges during this period. Events on Cheju-do, where members of the 9th Regiment tried to join the insurgents and the 11th Regiment's commanding officer was murdered by a member of his staff, indicated that leftists within its ranks were becoming increasingly bold. American CIC personnel, aided by South Korean intelligence officers loyal to Rhee, were ordered to purge the military of political undesirables.

---

[41] Andrew J. Birtle, *U.S. Army Counterinsurgency and Contingency Operations Doctrine, 1942–1976* (Washington, D.C.: U.S. Army Center of Military History, 2006), p. 94.

[42] "The Police and National Events, 1947–1948," in Army Forces in Korea, vol. 3, pt. 1C, p. 37.

[43] *South Korea: A Country Study*, p. 31.

[44] Millett, *The War for Korea*, p. 133.

As historian Brian Gibby explained, "With assistance from the U.S. Army's 971st Counterintelligence Detachment, Paik [Col. Paik Sun-Yup] managed to identify nearly all suspected subversives."[45] Aware that the net was closing in, Communist sympathizers in the South Korean Army decided to act.

On 19 October 1948, the 14th Regiment mutinied after receiving orders to replace the 11th Regiment on Cheju-do. A small group of mutineers killed thirty officers and persuaded six hundred to eight hundred of their comrades to switch sides. A Korean mechanized squadron, led by American Lts. Robert M. Shackleton and Ralph Bliss, defeated the main group of mutineers after prolonged fighting.[46] However, the mass defections "rocked the Rhee administration and the American mission in Seoul. . . . There was no question that the week-long fighting had been a major political event, despite . . . pitiful attempt[s] to cast the revolt as simply Communist terrorism with no roots in the population."[47]

Violent disaffection within the army was not confined to the 14th Regiment. On 2 November 1948, a senior noncommissioned officer of the 6th Regiment charged with inciting mutiny killed a Korean Military Police lieutenant sent to arrest him. The murderer gathered approximately one hundred enlisted men, who proceeded to kill seven more officers. South Korean Military Police rushed to the scene, only to encounter a deadly ambush. The survivors sought assistance from a nearby American unit, which confronted and disarmed the mutineers. About twenty rebel soldiers escaped, including the ringleader. Two more incidents involving the 6th Regiment occurred on 6 December 1948 and 30 January 1949. Both involved the deaths of several loyal officers and the escape of many mutineers.[48]

Encouraged by disarray within Rhee's armed forces and bolstered by statements of support from Joseph Stalin, Kim Il Sung began providing direct assistance to the southern insurgents.[49] Between 20 October 1948 and 28 March 1950, he dispatched ten guerrilla units, ranging in size from fifteen to seven hundred personnel, to South Korea. While security forces killed, captured, or dispersed almost all of the northerners, two groups managed to link up with southern dissidents in the city of Taedong.[50]

---

[45] Gibby, "Fighting in a Korean War: The American Advisory Missions from 1946–1953," p. 58.

[46] Ibid, p. 60.

[47] Millett, *The War for Korea*, p. 170.

[48] Military History of Korea, translated from Korean by Headquarters, U.S. Army Forces, Far East Military Intelligence Service Group, 5 Sep 52, pp. 55–56, Historical Research Collection, CMH.

[49] Kathryn Weathersby, *Soviet Aims in Korea and the Origins of the Korean War, 1945–1950: New Evidence from Russian Archives*, Working Paper 8, Cold War International History Project (Washington, D.C.: Woodrow Wilson International Center for Scholars, 1993), pp. 23–24.

[50] Military History of Korea, pp. 46–48.

Rhee reacted to the northern infiltration by ordering his commanders to conduct mass counterinsurgency sweeps. The gravity of the situation was highlighted by Rhee's acquiescence to American suggestions to assign the most capable officers to oversee the sweeps. As historian Alan Millett observed, "Desperate for some military success, the president permitted the American advisers more latitude in selecting commanders and designing operations, which allowed the professional officers of the first class to conduct field operations in cooperation with their American counterparts and stay-behind elements of USAFIK [United States Army Forces in Korea]."[51]

While American advisers and their Korean counterparts found themselves "in constant conflict over proper counterinsurgency methods, . . . out of this tension came a meld of American methods and the techniques of suppression that the Japanese had developed in Manchuria, for combating guerrillas in cold-weather, mountainous terrain, implemented by Korean officers who had served the Japanese (often in Manchuria)."[52] The hybrid tactics resulted in several significant coups. Kim Yong Kwan, leader of the People's Committee on Cheju-do, was killed on 20 April 1949.[53] During that same period, the leader of the Yosu Rebellion, Lt. Kim Chi-hoe, died of wounds following a skirmish with loyal troops.[54]

By late 1949, the flagging efforts of the southern dissidents strengthened Kim Il Sung's conviction that a full-scale invasion would be necessary to unify the Korean peninsula.[55] When the North Korean attack began on 25 June 1950, the American advisory group consisted of 183 officers and 286 enlisted men.[56] South Korean losses in personnel and equipment were so great during the first month of the conflict that three divisions were inactivated. Faced with a disintegrating southern army, American advisers assumed responsibilities well beyond the norm. Counterinsurgency receded into the background as Korean and U.S. forces concentrated on surviving.

Ignoring the southern insurgency proved easier said than done. The North Koreans considered the dissident groups as potentially valuable allies. Likewise, the insurgents intended to make the most of the changing military situation. As Communist conventional forces surged southward, "guerrillas became a constant threat to the rear areas of ROK and Allied troops fighting

---

[51] Millett, *The War for Korea*, p. 202.

[52] Ibid., p. 286.

[53] Military History of Korea, p. 49.

[54] Ronald H. Spector, *In the Ruins of Empire: The Japanese Surrender and the Battle for Postwar Asia* (New York: Random House, 2007), p. 204.

[55] There is evidence that Kim Il Sung had contemplated a full-scale invasion prior to this period. See Bruce Cumings, *The Origins of the Korean War, Volume II, The Roaring of the Cataract, 1947–1950* (Princeton, N.J.: Princeton University Press, 1990), p. 284.

[56] "Narrative Account of Events in South Korea from 25 June through 1 July 1950," in Military History of Korea, p. 1.

their bitter withdrawal to the Pusan Perimeter."[57] The invaders installed a provisional government, staffed in many instances with southern insurgents, in the wake of their victorious advance.

When the UN forces broke out of Pusan in September 1950, the changing fortunes of war forced leftist insurgents in the provisional Communist government to flee. Although the northern military had occupied South Korean territory for a relatively brief period, its oppressive actions convinced many southerners that President Rhee seemed preferable to permanent Communist rule. As a result, South Koreans eagerly stepped forward to identify insurgents as soon as UN forces liberated their villages. An American intelligence report, for example, listed no less than twenty-six people in the village of Chogye as collaborators and dissidents.[58] Most of the unmasked insurgents were arrested and shot by the Korean National Police.

Although Korean security forces eliminated thousands of collaborators and guerrillas during the fall of 1950, the rear area threat actually swelled as stay-behinds from five North Korean divisions began harassing UN forces.[59] The northerners absorbed, operated alongside, or concluded informal alliances with surviving local dissidents. An American officer, recording the result of the joining of irregular forces, noted, "For seventeen months following the outbreak of the war, the southwest guerrillas blazed a trail of destroyed farms, shattered communications routes, and defiance of the provincial governments. Mobilized against them was the National Police, organized on full war footing and fighting battles which at times compared in size with those on the main battle front."[60]

The Korean National Police had always depended on military assistance to battle insurgency. By November 1950, however, the bulk of the prewar Korean Army had been dispersed or destroyed. By default, the mission of assisting the Korean National Police temporarily fell on American and UN forces. The Americans launched their first major counterinsurgency operation on 18 January 1951, when the 1st Marine Division, augmented by two battalions of South Korean marines, deployed to protect the supply routes for UN forces located south of Seoul.[61] Southern insurgents and their North Korean allies followed the traditional guerrilla pattern of dispersing when

[57] Conduct of Anti-Guerrilla Operations in Southwest Korea by Task Force Paik, 2 December 1951—8 February 1952, p. 3, Records of General Headquarters, Far East Command, Supreme Commander Allied Powers, and United Nations Command, Record Group (RG) 554, National Archives and Records Administration, College Park, Md. (hereafter cited as NACP).

[58] An. 3 to Periodic Intelligence Rpt 32, 2d Inf Div, 302–2.1, G–2 Periodic Reports, Sep–Oct 50, Records of the Adjutant General, RG 407, NACP.

[59] These units included the 2d, 4th, 5th, 7th, and 10th Divisions (History of the North Korean Army, sec. V, pp. 54—69, Historical Manuscript Collection, CMH).

[60] Conduct of Anti-Guerrilla Operations in Southwest Korea by Task Force Paik, 2 December 1951—8 February 1952, p. 3.

[61] Lynn Montross et al., *The East–Central Front, U.S. Marine Operations in Korea, 1950–1953* (Washington, D.C.: Historical Branch, HQ, U.S. Marine Corps, 1962), 4:3.

confronted by vastly superior military force. For twenty-seven days, U.S. and South Korean marines chased down guerrillas, *North Korean People's Army* (*NKPA*) stragglers, and rumors. In return for 184 friendly losses, the marines killed 120 enemy and took 184 prisoners.[62]

The experience of the 10th Battalion, Philippine Expeditionary Force to Korea, provides an example of smaller scale UN participation. Possessing some experience in fighting insurgents, the 10th received orders to eliminate insurgent bands threatening supply routes between Seoul and the ports of Masan and Pusan. The battalion's operational area covered more than eight hundred square miles and harbored about three thousand guerrillas. The 10th spent the initial months of its tour in Korea conducting numerous sweeps for northern stay-behinds and southern dissidents, until it finally received a front-line assignment with the 187th Airborne Regimental Combat Team.[63]

Not surprisingly, American and UN commanders in Korea found it extremely difficult to conduct successful counterinsurgency operations. Since their troops remained only long enough in a given area to hunt for guerrillas, local residents were hesitant to pass along any information about insurgent bands. Their fear of reprisal after UN troops left proved greater than any sense of civic duty. The total lack of useful intelligence was only partially offset by UN military liaison cells collocated with special security elements of the Korean National Police.[64]

The long-term solution lay in the United Nations Command providing additional training and equipment to the South Koreans. Talks between General James H. Van Fleet, commander of the UN forces in Korea, and Maj. Gen. Lee Chong Chan, South Korean chief of staff, resulted in an agreement to expand the South Korean Army to 10 infantry divisions, 22 artillery battalions, 3 armor battalions, 13 security battalions, and 10 separate infantry regiments.[65] With security battalions and separate regiments, which were not earmarked for front-line operations, constituting one-third of the South Korean infantry strength, the rear threat was taken very seriously.

Although UN counterinsurgency sweeps never gained lasting success, the increased number of American advisers assigned to the expanding South Korean Army had a significant impact on South Korean counterinsurgency operations. As historian Andy Birtle observed,

---

[62] Ibid., p. 57.

[63] Art Villasanta, The Glory of Our Fathers: The Philippine Expeditionary Force to Korea, 2000, Historians files, CMH.

[64] Intelligence and Counterintelligence Problems During the Korean War, Military History Section, Headquarters United States Army Forces and Eighth United States Army, n.d., p. 26, Historical Manuscript Collection, CMH.

[65] Kenneth W. Meyers, United States Advisory Group to the Republic of Korea, pt. 4, KMAG's Wartime Experiences, 11 July 1951 to 27 July 1953, n.d., Office of the Military History Officer, U.S. Army, Japan, p. 43, Historical Manuscript Collection, CMH.

> The widening war increased American influence over the conduct of South Korean counterinsurgency operations, as Rhee subordinated the ROK Army to American command and the United States deployed additional advisers. Several of the new U.S. personnel had had significant counterguerrilla experience prior to deploying to Korea. Foremost among these was General [James] Van Fleet, who assumed control over allied ground forces in Korea in the spring of 1951.[66]

December 1951 witnessed the Korean Army's return to deliberate counterinsurgency operations. Task Force Paik—led by Lt. Gen. Paik Sun Yup and consisted of a combined U.S.-Korean headquarters, the Capital and 8th Infantry Divisions, Korean police and security battalions, and American support units—embarked on a three-month operation against guerrillas in southwestern Korea. General Paik later commented, "I must say that it was risky to withdraw two full divisions from the front and employ them in rear-area combat operations during a shooting war. Indeed, our willingness to accept that risk suggests the level of threat the communist guerrillas posed to the state."[67]

Paik's staff included an American team composed of sixty experts at operational liaison, communications, air-ground liaison, engineering, and psychological operations. His senior U.S. adviser, Lt. Col. William A. Dodds, had accumulated considerable counterguerrilla experience in Greece.[68] A total of 530 American officers and men, including attached engineer and transportation units, provided logistical aid to Task Force Paik.[69] The Americans also furnished six L–19 liaison aircraft and one helicopter.[70]

From the onset, Paik prohibited his soldiers from taking food, burning homes, and firing at individuals not actively resisting. In return, South Korean civilians who had refused to help soon began providing Paik's troops with information about the guerrillas. The task force's after action report remarked, "Not the least reason for this change in attitude was the excellent conduct of the Task Force troops, which in turn was the result of the emphasis placed by the Task Force commander upon good behavior as part of good public relations."[71]

In a four-phased operation, Task Force Paik killed, wounded, or captured 19,779 guerrillas. The South Koreans seized a total of 3,239 weapons, including mortars, recoilless rifles, machine guns, rifles, and pistols.[72] South Korean and U.S. Air Force tactical aircraft, rather than field artillery, provided the necessary fire support. General Van Fleet, who "learned that the infantry

---

[66] Birtle, *U.S. Army Counterinsurgency and Contingency Operations Doctrine, 1942–1976*, pp. 100–101.

[67] Paik Sun Yup, *From Pusan to Panmunjom* (New York: Brassey's, 1992), p. 182.

[68] Ibid., p. 183.

[69] Conduct of Anti-Guerrilla Operations in Southwest Korea by Task Force Paik, 2 December 1951–8 February 1952, p. 22.

[70] Ibid., pp. 12–13.

[71] Ibid., p. 20.

[72] Ibid., p. 10. The majority of the weapons were Russian-made.

would not execute a final assault on a summit defended by die-hard guerillas if artillery support was available," insisted upon this tactic.[73] While Paik's dramatic success did not completely eliminate the United Nations Command's rear-area security problems, he demonstrated that it was possible to curb the activities of southern insurgents and northern irregulars.

A massive resurgence of political violence, originating from a familiar source, appeared imminent in South Korea during the spring of 1952. President Rhee's attempts to amend the constitution to ensure that he would be reelected by popular vote rather than chosen by the assembly encountered stubborn opposition from political foes. On 24 May, an exasperated Rhee declared martial law in Pusan and placed a number of his opponents under arrest.[74] Concerned that South Koreans would begin venting their anger against UN personnel and installations, Van Fleet sent an American rifle battalion to Taegu as a reserve force to respond to rioting. He also readied an American regimental combat team in Japan to deploy to South Korea to perform security duties if the tense political situation adversely affected ongoing UN logistics operations.[75] Under considerable American pressure, to include intervention by President Harry S. Truman, Rhee and the National Assembly negotiated a face-saving compromise averting the possibility of renewed internal rebellion.

Conclusions

Summers' use of Korea as an example where the U.S. military "successfully" separated counterinsurgency from conventional combat reflects an attempt to intellectually come to grips with the outcome of the Vietnam War rather than objective analysis. Colonel Alnwick offers a more nuanced assessment by noting, "The United States never had a choice between counterinsurgency and national building on one hand and conventional warfare on the other, to have any chance of winning or even avoiding defeat, we had to devote equal energy to both tasks."[76]

Separating counterinsurgency and conventional warfare into two distinct categories also suggests that there is no link between both types of conflict. In doing so, Colonel Summers overlooks historical examples, such as the Communist Chinese rebellion in the late 1940s, where an insurgent force transformed into a conventional army. Unlike Alnwick, who argues that "two decades of relentless guerrilla warfare" set the stage for South Vietnam's defeat, Summers' statement dismisses the possibility of an insurgency establishing the preconditions for conventional military operations.[77]

---

[73] Paik, *From Pusan to Panmunjom*, p. 184.

[74] Walter G. Hermes, *Truce Tent and Fighting Front* (Washington, D.C.: U.S. Army Center of Military History, 1988), p. 345.

[75] Ibid., p. 346.

[76] Alnwick, *Strategic Choice, National Will, and the Vietnam Experience*, p. 136.

[77] Ibid.

Summers' aversion to nation building, which reflected the attitude of many Army officers after Vietnam, overlooks the root causes for an insurgency. In Korea, for example, the American Military Government opted for a low-cost solution that returned Korean members of the Japanese Colonial Police to their former jobs. The brutal behavior of the police, who openly supported right-wing political factions, fueled rather than inhibited the rise of South Korean insurgent groups. Allotting sufficient resources at the onset to American nation-building efforts in Korea, at least in this case, could have produced a police force capable of solving rather than adding to the internal security challenges facing the American Military Government.

By reinforcing the legitimacy of the indigenous government in the eyes of its own population, nation building, even at the most localized level, forms a key component of any counterinsurgency campaign. An American report published in 1951 stated, "Communist forces will find it hard to grow or even exist among people who are well fed, well housed, well clothed, and gainfully employed. On the other hand, it is useless to believe in the ultimate success of any [purely] military [counterinsurgency] operation if [poor social] conditions continue to foster political or economic intent."[78] General Paik demonstrated a keen understanding of this crucial relationship.

Summers' comparison wrongly assumes the presence of effective indigenous security forces to wage the counterinsurgency fight while American combat units focus on conventional warfare. Precisely the opposite situation occurred during the Autumn Harvest Rebellion in 1946, when Communist sympathizers succeeded in undermining the constabulary. Although the South Korean Army gained many successes in 1948–1949, it withdrew from the counterinsurgency fight when it respectively suffered crippling losses at the hands of the North Koreans in 1950 and the Chinese in 1951. American and UN military forces, in close cooperation with Korean National Police, assumed responsibility for rear-area security and counterinsurgency until the reconstitution of the South Korean Army in late 1951.

Thus American involvement in South Korean counterinsurgency operations spanned the entire spectrum, from "hands off" to intimate participation. Although U.S. commanders would have undoubtedly preferred to pass responsibility for counterinsurgency operations to the South Koreans, significant changes to the strategic and political situation frequently prevented them from doing so. While Summers' choice of a "Korea counterinsurgency model" remains highly debatable, his observations serve as an indicator of how our Army's experience in Korea, rightly or wrongly, influenced American decisions during Vietnam.

---

[78] Birtle, *U.S. Army Counterinsurgency and Contingency Operations Doctrine, 1942–1976*, p. 89.

Lessons Learned and Relearned: Gun Trucks on Route 19 in Vietnam

Ted Ballard

The war in Iraq is, as the Vietnam War was, a war without front lines. The relatively free movement of enemy insurgents throughout the country makes U.S. Army supply lines vulnerable and invites attacks on vehicle convoys. This proved especially true early in the Iraq War, when U.S. truck convoys lacked sufficient internal or organic security to deter or defend against aggressive attacks. In these attacks, the enemy often employed rocket-propelled grenades and heavy machine guns, as well as the AK47 assault rifle. Military police (MP) units often provided the only security for truck convoys, but in most cases MP units did not possess sufficient organic firepower to withstand a determined enemy attack. For example, in March 2003, Iraqis ambushed a U.S. Army truck convoy and made headlines by killing eleven soldiers and wounding and capturing seven, including Pfc. Jessica Lynch. Only the arrival of a nearby tactical unit prevented the loss of additional personnel.[1] This ambush was followed by similar attacks on U.S. convoys in Iraq, with loss of life and vehicles.

The constant threat of attacks on convoys led some Army transportation units in Iraq to obtain local scrap steel and place it on five-ton cargo trucks. They also placed machine guns in the vehicle beds and reassigned drivers as gunners.[2] Since these gun trucks formed an integral part of the convoy and could react instantly to enemy attacks, they had an operational effectiveness unmatched by the available external security assets. As more transportation units in Iraq began to "harden" vehicles for convoy security, the Army made armor-plate kits available through official channels. In 2004, the Lawrence Livermore National Laboratory designed modular, heavy-duty armor kits to convert cargo trucks into convoy protection vehicles. By 2005, thirty-one cargo trucks, using the armor-plate kit, beefed up convoy security in Iraq.[3] The innovation of the cargo truck converted to an armored gun truck soon became a familiar sight accompanying convoys on the roads of Iraq.

---

[1] Attack on the 507th Maintenance Company, 23 March 2003, An Nasiriya, Iraq, U.S. Army Training and Doctrine Command, Fort Monroe, Va.

[2] Capt. Daniel T. Rossi, "The Logistics Convoy: A Combat Operation," *Army Logistician* 37 (January–February 2005).

[3] Public Affairs News Release, Lawrence Livermore National Laboratory, 21 Jul 2005.

This author served with the 523d Transportation Company in Vietnam, and the convoy security problem and its solution in Iraq gave me a sense of déjà vu. Transportation personnel in Iraq confronted a convoy security problem and, like a previous generation of Army truckers in Vietnam, arrived at the same solution.

During the deployment of tactical units to Vietnam in mid-1965, the Army stationed most highway transport units at or near the major port areas. They provided port and beach clearance and local and line haul from ports at Saigon, Cam Ranh Bay, and Qui Nhon. In October 1966, the 8th Transportation Group arrived in Vietnam and was assigned to the coastal and fishing center at the port of Qui Nhon. The 8th Group assumed operational command of three trucking battalions: the 27th and 54th, near Qui Nhon, and the 124th, more than one hundred miles west on Route 19 at the Central Highland town of Pleiku. The mission of these battalions was to provide motor transport to tactical forces in the II Corps Tactical Zone. This article focuses on the 54th Transportation Battalion and its experiences on Route 19. However, the innovations mentioned in this work took place in surrounding transportation battalions as well.

The 54th Transportation Battalion compound (named Camp Addison after a trucker killed in action) was located sixteen miles northwest of Qui Nhon, along Route 19.[4] The 54th engaged in line haul operations from the port at Qui Nhon to various combat units and stations on Route 19 to the west or Route 1, which ran parallel to the coast. The battalion initially included three light truck companies—the 512th, 523d, and 669th—and one medium company, the 563d.[5] Another light truck company, the 666th, joined the battalion in August 1967.[6] The light companies were equipped with M54 five-ton cargo trucks, and the medium company with M52 tractors with twelve-ton trailers.[7] Convoys hauled a variety of supplies to tactical units, anything from beer to artillery shells and from building materials to 500-gallon rubber bladders filled with JP–4 aviation fuel.

The battalion divided convoy operations into daytime and nighttime trips. During the night, drivers moved empty trucks from the Camp Addison motor pool to various compounds in and around Qui Nhon, had the trucks loaded, and then returned them to the camp motor pool. Known as night shuttles, these convoys might contain one, two, or more trucks, and some drivers might make more than one trip to Qui Nhon and back during the night. By morning, all trucks were loaded and ready for day drivers.

---

[4] HQ, 54th Transportation Bn (Truck), Annual Supplement to Unit History, 1 June 1966 to 31 December 1967, 28 Mar 67.

[5] Ibid.

[6] HQ, 54th Transportation Bn (Truck), Operational Report for the Quarterly Period Ending 31 October 1967, RCS CSFOR-65, 1 Nov 67.

[7] HQ, 54th Transportation Bn (Truck), Operational Report for the Quarterly Period Ending 30 April 1967, RCS CSFOR-65, 13 May 67.

Daytime convoys might have anywhere from 100 to 230 vehicles, with the column of trucks stretching for miles.[8] Because most of Route 19 was unpaved and heavily potholed, the battalion's official standard operating procedure (SOP) required that convoys obey designated speed limits and reduce speeds commensurate with road, weather, and traffic conditions. The SOP set convoy speeds at twenty-five miles per hour (mph), not to exceed thirty-five mph, and reduced speed through villages to approximately fifteen mph.[9]

Route 19 served as the main ground supply route from the port at Qui Nhon to the town of Pleiku in the Central Highlands, the longest line haul route in Vietnam. The route west from Camp Addison traversed one hundred miles of winding, potholed, mostly unpaved roadway, crowded with civilian traffic. The road was bordered in some places by wide-open fields or rice paddies and in other places by heavy jungle. It crossed bridges about every three miles, most of which had been destroyed and replaced by temporary bridging. About fifty miles west of Qui Nhon, Route 19 left the coastal plains and climbed five thousand feet through two mountain passes. The first, the An Khe Pass, contained several switchbacks and a nasty hairpin turn, nicknamed the "Devil's Hairpin" by the truckers. Often, near the top of the pass, the trucks would literally enter the clouds, forcing the drivers to decrease the distance between vehicles in order to keep visual contact with the vehicle in front. Stationed near the top of the pass was the 1st Cavalry Division with its fleet of almost five hundred helicopters.

From An Khe, Route 19 continued westward to the infamous Mang Yang Pass. In 1954, along this same highway, the Viet Minh annihilated a French military convoy, Group Mobile 100, during the French-Indochina War. Near the top of the pass, a small stone marker commemorating the battle still stands. After reaching the top of the pass, Route 19 continued approximately twenty more miles to the camps of the 4th Infantry Division at Pleiku. There, the convoy dispersed into smaller groups of vehicles, depending on the material on each truck, and moved to various supply areas for unloading. The empty trucks then reassembled for the return trip.

Poor road conditions and speed restrictions made the one-way, 100-mile-plus trip from Camp Addison to Pleiku take about five hours. Unloading the large number of trucks took another three hours. The 200-mile-plus round trip might take about thirteen hours, but often the poor road and weather conditions lengthened that time. Since these convoys moved seven days a week, and a shortage of drivers required every man to remain on duty, the hours for the drivers were grueling, with some drivers on the road nonstop as many as fifteen to eighteen hours at a time.[10]

---

[8] HQ, 54th Transportation Bn (Truck), Operational Report for the Quarterly Period Ending 30 April 1967, RCS CSFOR-65, 13 May 67.

[9] Col Joe O. Bellino, 8th Transportation Group, Sep 1967–Sep 1968, n.d.

[10] HQ, 54th Transportation Bn (Truck), Operational Report for the Quarterly Period Ending 30 April 1967, RCS CSFOR-65, 13 May 67.

A convoy commander, usually a platoon leader in an M151 quarter-ton truck equipped with a PRC–25 battery-operated radio, accompanied each convoy. Since the PRC–25 had a range of only a few miles, the departing convoy was soon out of contact with battalion headquarters at Camp Addison. As the convoy passed near tactical units stationed along the highway, the convoy commander could use his radio to alert tactical units that the convoy was in the area or to call them for assistance if necessary. Unfortunately, considerable maintenance problems plagued the PRC–25 radios, which often malfunctioned.[11]

Tactical units along Route 19 between Qui Nhon and Pleiku provided limited convoy security. But some stretches of highway had no tactical forces available. From Qui Nhon to the foot of the An Khe Pass, a South Korean infantry division had responsibility for guarding the highway's many bridges. No tactical units occupied the An Khe Pass itself because of the difficult terrain. The 1st Cavalry Division provided security from the top of the pass to the base of the Mang Yang Pass, but again, no tactical units covered that pass. Just beyond the top of this pass, the elements of the 4th Infantry Division picked up road security to Pleiku. Tactical units only provided convoy security when priorities permitted. Since U.S. forces in the Central Highlands often engaged in combat missions, they just as often failed to provide the desired degree of support for convoys.

Although not ideal, the security system for the daily convoys in place in the early part of 1967 seemed to work. Up to that time, little significant enemy activity directly involved 54th Battalion convoys. By April, only two sniping incidents, which wounded drivers, involved 54th convoys traveling Route19.[12] Still, the battalion headquarters remained concerned for convoy security. By May, the battalion commander requested several .50-caliber machine guns with ring mounts. He intended to mount them on selected cargo trucks interspersed within the convoys. In the event of ambush, the drivers of these armed trucks would enter the ring mount and return fire.[13] To protect themselves from possible mines, the 8th Group drivers also were instructed to add sandbags to the floors of their trucks. This lessened the possibility of mine injury, but the increased floor height often made working accelerator and brake pedals difficult. And, to increase command and control, the 8th Group reduced the number of vehicles in a single convoy to one hundred or fewer trucks.[14]

As fall 1967 approached, the convoys continued their daily trips on Route 19 and Route 1, and enemy activity remained at a low level. The requested

---

[11] HQ, 54th Transportation Bn (Truck), Operational Report of the 54th Transportation Battalion (Truck) WFR6AA, for the Period Ending 31 July 1968, 6SFOR-65 (R-1), 1 Aug 68.

[12] HQ, 54th Transportation Bn (Truck), Operational Report for the Quarterly Period Ending 30 April 1967, RCS CSFOR-65, 13 May 67.

[13] Ibid.

[14] HQ, 54th Transportation Bn (Truck), Operational Report for the Quarterly Period Ending 31 October 1967, RCS CSFOR-65, 1 Nov 67.

.50-caliber machine guns and ring mounts had yet to arrive at the 54th, but there seemed to be no urgency in arming the unit's trucks. An occasional mine incident or lone sniper taking a potshot at a convoy was more of a nuisance than a perceived threat of anything more serious.

All that changed on 2 September 1967. About 7:00 p.m., an eastbound 54th convoy of thirty-seven empty vehicles was returning from Pleiku. As the column approached the An Khe Pass, a 57-mm. recoilless rifle round struck the lead M151, killing one passenger and wounding the driver and machine gunner. Surprised by the attack, the drivers in the following cargo trucks halted and dismounted. A loaded, 5,000-gallon fuel tanker had joined the convoy near Pleiku, and enemy rounds soon set the vehicle on fire, blocking the road. Drivers behind the tanker also jumped from their vehicles. In the confusion of the attack, none of the drivers seemed to know what to do. With the convoy now spread out along the highway and with little or no communication with their officers and noncommissioned officers, the dismounted drivers were simply a group of individual soldiers without a plan of action. As some drivers returned fire, the enemy, at estimated company strength, charged in among the trucks firing small arms and throwing hand grenades. Local tactical security units from the nearest checkpoint heard the firing and arrived within fifteen minutes, but by then the enemy had withdrawn under cover of darkness. The ambush had been a great success for the enemy. In less than fifteen minutes, he killed seven drivers, wounded seventeen, and damaged or destroyed thirty of the thirty-seven vehicles.[15] The convoy had proved an easy target. For the enemy, this ambush began a series of attacks on the convoys intended to shut down the supply line on Route19 for the five months before the Tet offensive of 1968.

In response to the September attack, the commanders of the 54th Battalion and its higher headquarters, the 8th Transportation Group, took several steps to counter any further attacks. They realized that the ambushed convoy had been on the road too late in the evening, which had allowed its attackers to escape in the darkness before nearby tactical forces arrived. Consequently, the commanders closed Route 19 to traffic traveling east of An Khe at 5:00 p.m., instead of the usual 7:00 p.m.[16] Since the enemy had launched his attack from concealed positions in the brush close to the road, engineers using bulldozers began clearing away vegetation one thousand meters from either side of the highway near the ambush site.[17]

The convoys also adopted a change in doctrine—the Hardened Convoy Concept. This was no more than utilizing special communications and armor-plated equipment for control purposes and firepower. Battalion convoys would also run in serials, with ten vehicles per serial, spaced approximately

[15] Bellino, 8th Transportation Group.

[16] Ibid.

[17] Richard E. Killblane, *Circle the Wagons: The History of US Army Convoy Security* (Fort Leavenworth, Kans.: Combat Studies Institute Press, 2005), p. 29.

five minutes apart. At the head of each serial was an M151 gun jeep equipped with a radio and armed with a pedestal-mounted machine gun, with a "gun truck" immediately behind it.[18]

The gun trucks, modified M35 2½-ton cargo vehicles, were armor-plated and sandbagged; staffed with a driver, an assistant driver, and a gun crew in the bed of the vehicle; and carried two M60 machine guns. Initially, the machine guns were simply carried on the floor of the bed of the truck; later, they were placed on pedestal mounts. In addition to a grenade launcher and machine guns, each crew member was armed with his issued M14 rifle. The Americans employed 2½-ton vehicles as gun trucks because every truck company had at least one of these vehicles, which it used for a variety of administrative purposes around camp. Second, converting the 2½-ton trucks for convoy security would not deplete the battalion's fleet of 5-ton task vehicles for daily convoys. The battalion quickly began locating local scrap steel and placing it on the sides of the 2½-ton trucks. Soon precut steel plate armor kits arrived from the States, and the truckers quickly placed them on the 2½-ton gun trucks. These kits included two quarter-inch steel doorplates, three quarter-inch plates for the bed of the truck, and a windshield cover. To provide additional firepower, the units equipped a few of the 2½-ton trucks with M55 quad .50-caliber machine guns. The M55 consisted of four .50-caliber machine guns, operated electronically, with a single gunner firing all four guns in tandem. The M55 could fire 450 to 550 rounds per minute. Soon the 54th Battalion had converted a dozen 2½-ton cargo trucks into gun trucks.[19]

The 54th's convoys, now accompanied by the gun trucks, continued to make the daily haul of supplies west to An Khe and Pleiku. Until late November, they reported only a few minor sniper and mining incidents. None halted the daily movement of supplies to U.S. tactical units in the Highlands.

On the morning of 24 November 1967, the enemy again launched a major ambush of a 54th convoy on Route19, west of An Khe. On that day, the westbound morning convoy included forty-three 5-ton cargo trucks, fifteen 2½-ton trucks of another battalion, a maintenance truck, six 2½-ton gun trucks, and three gun jeeps. The convoy was divided into six serials of about ten task vehicles per serial, with one gun truck leading each serial. About 10:00 a.m., a rocket-propelled grenade struck and destroyed the gun truck leading the first serial. Simultaneously, a mine destroyed the lead cargo truck. Other truck drivers attempted to escape the kill zone, but small arms fire or mines knocked out trucks and blocked the road. Several 5,000-gallon fuel tankers in the first serial burst into flames. Only one truck managed to escape the inferno. Another rocket-propelled grenade struck the gun truck leading the second serial, wounding the driver and causing the truck to careen off the road and flip over, crushing to death one of the crew. Enemy fire hit cargo trucks loaded

---

[18] Bellino, 8th Transportation Group.

[19] Ibid.; Killblane, *Circle the Wagons*, p. 32.

with artillery ammunition, which began to explode. When one ammunition truck detonated, the explosion destroyed one of the gun trucks directly behind it. Rocket-propelled grenades damaged and knocked out two more gun trucks, but the remaining two armored vehicles in the last two serials moved up to the kill zone and began to lay down a suppressive fire on the attackers.

When the first gun truck was struck, the convoy commander had immediately radioed the call sign, "Contact," alerting a nearby checkpoint. Within twenty minutes, tanks and armored personnel carriers arrived on the scene, but the enemy had already withdrawn. The attack killed three truckers, wounded seventeen, and destroyed or damaged ten cargo trucks and four of the six gun trucks. Although the convoy suffered casualties and vehicle damage, the firepower of the remaining two gun trucks, along with that of individual truckers, had prevented a repeat of the September ambush. The enemy suffered forty-one killed and four wounded and captured.[20]

Less than two weeks later, the enemy tried again to destroy a 54th convoy. On 4 December, a company-size enemy force ambushed an eastbound convoy returning from Pleiku. This convoy consisted of six serials, totaling fifty-eight 5-ton trucks and eleven 2½-ton trucks. A gun truck led each serial. The attack began when a rocket-propelled grenade struck the lead gun truck. It killed the driver, wounded the crew, and halted the first serial. As the enemy made several attempts to reach the cargo trucks, the remaining five gun trucks drove into the 3,000-meter-long kill zone and opened suppressive fire on the enemy. Rockets struck and disabled two more of the gun trucks, wounding the crews. The crews of all gun trucks, including the disabled vehicles, continued to deliver a heavy fire on the enemy. Calls for assistance to nearby tactical units by the convoy commander resulted in the arrival of several helicopter gunships, but the enemy had withdrawn, leaving behind thirteen dead and one wounded. U.S forces suffered one trucker killed and six wounded. The enemy destroyed one gun truck and slightly damaged four cargo trucks. In this attack, gun trucks had suppressed enemy fire and driven off the attack. The actions of the gun trucks had demonstrated that they could reduce convoy losses significantly while inflicting increased damage on the enemy. In the ambushes in late 1967 and early 1968, the enemy had opened fire on the first vehicles in the convoy, the gun trucks. However, in subsequent ambushes, the enemy modified his tactics by allowing the gun trucks to pass through the kill zone and waiting to attack the cargo trucks.[21]

During 1968, the 54th Transportation Battalion continued to refine the gun truck concept. In a major change, it began to armor-plate five-ton cargo trucks and convert them into gun trucks.[22] The "deuce-and-a half" lacked the

[20] Bellino, 8th Transportation Group.

[21] HQ, 8th Transportation Gp (Motor Transport), Operational Report for Quarterly Period Ending 31 January 1968, RCS CSFOR-65, 15 Feb 68.

[22] HQ, 54th Transportation Bn (Truck), Operational Report of the 54th Transportation Battalion (Truck) WFR6AA, for the Period Ending 31 October 1968, 6SFOR-65 (R-1), 3 Nov 68.

horsepower to handle the additional weight of armor, sandbags, and weapons. Also the 2½-ton trucks were frequently in the motor pool for repairs because the extra weight put too much strain on the vehicles. Although superior to the 2½-ton truck, the widespread use of the 5-ton as a convoy security vehicle reduced the number of cargo trucks available for line haul operations.

Another change in gun-truck design modified the armor on some gun trucks by adding double-walled "box" structures to the beds. In theory, the double-wall acted as a protection against enemy rockets. Supposedly, incoming rockets would detonate on contact with the outer wall, causing the resulting fragments to bounce harmlessly off the inner wall, never reaching the crew. Soldiers kept extra weapons, tools, extra wheels and tires, water, and a fire extinguisher between the walls. Thus, the gun trucks served not only as security vehicles but also as maintenance trucks, capable of protecting and restarting disabled vehicles on potentially dangerous sections of highway.[23] Some transportation companies, not always able to obtain armor plate, had the hulls of armored personnel carriers stripped and placed in the beds of gun trucks. However, the increased height of these gun trucks resulted in their being top-heavy and difficult to control on the highway, and the extra weight of the armored hull increased maintenance problems.

Crews often replaced the M60 machine gun on gun trucks with the heavier .50-caliber M2 machine gun, either alone or on multiple mounts. In some cases, they added multibarreled 7.62-mm. miniguns. These miniguns, designed to provide a high rate of fire, sprayed out either 2,000 or 4,000 rounds per minute.

Another gun-truck development involved communication. At first, gun trucks did not have radios and, therefore, operated on their own orders or under verbal orders of the convoy commander, which could be difficult during an ambush. The battalion attempted to remedy this by placing PRC–25 radios in each gun truck, which, again in theory, enabled gun trucks to maintain constant communication with the convoy commander and other gun trucks. This allowed all gun trucks to be alerted instantly to mines and ambushes and to coordinate responses. However, the PRC–25 radios continued to have maintenance problems—of the ten radios in the battalion, only five usually operated at any given time.[24]

The gun-truck crews had by now become proficient in protecting their fellow drivers on the highways. When then enemy attempted to ambush and stop a convoy on Route 19 in August 1968, two gun trucks halted in the

---

[23] HQ, 54th Transportation Bn (Truck), Operational Report of the 54th Transportation Battalion (Truck) WFR6AA, for the Period Ending 30 April 1970, (RCS CSFOR-65) (R-1), 2 May 70.

[24] HQ, 54th Transportation Bn (Truck), Operational Report of the 54th Transportation Battalion (Truck) WFR6AA, for the Period Ending 31 July 1968, 6SFOR-65 (R-1), 1 Aug 68.

kill zone and drove the enemy off. Although five cargo vehicle drivers were wounded, the convoy never halted.[25]

The gun-truck crews consisted, for the most part, of truckers and maintenance personnel who had been assigned or had volunteered for convoy security duty. Even as casualties among the crews mounted, morale remained high, as the men were bound by wartime esprit de corps. As the Hardened Convoy Concept spread to other transportation battalions, many of the gun-truck crews donned shoulder insignia and pocket tabs, indicating their roles as drivers and gunners. The gun-truck crews also began to paint their vehicles an intimidating black, with bright red, orange, yellow, or white trim. They painted colorful names on their vehicles, reminiscent of the American bombers of World War II. Names often reflected the popular culture of the time. For example, "Iron Butterfly" was named after a rock band, "The Untouchables" and "The Bounty Hunter" after popular TV shows, "Road Runner" for the animated cartoon character, "Godzilla" after the movie monster, and "Eve of Destruction" for a lyric in a popular song. Other names included "Ho Chi's Hearse," "VC Undertaker," "Highland Raiders," and "Old Ironsides."

Awards such as the Silver Star medal and Bronze Star medal with the "V," for valor, device, as well as Purple Hearts, became common among the gun-truck crews as they continued to place themselves between the truckers and the enemy. The personal bravery and fighting spirit of the gun-truck crews, along with their quick reaction to ambush situations, saved the lives of many truckers.

Specs. Dallas Mullins of the 444th Transportation Company and Larry G. Dahl of the 359th Transportation Company exemplified this courage. When the enemy wounded the driver of Mullins' gun truck during a highway ambush, the vehicle stalled in the center of the enemy kill zone and was subjected to intense small arms fire. Even though Mullins was also wounded (twice in the arm and once in the leg), he came to the aid of the wounded driver and maneuvered the truck out of the line of fire. During a 1971 ambush on Route 19 near An Khe, Dahl was a gunner on a gun truck called to assist other gun trucks involved in a convoy ambush. After driving off the enemy force, and as the gun trucks left the area, an enemy soldier tossed a hand grenade in the back of Dahl's truck. Dahl called a warning to his companions and threw himself directly onto the grenade. For their unselfish acts, Mullins earned the Silver Star, and Dahl, posthumously, the Medal of Honor.[26]

From late 1967 to the withdrawal of U.S. combat forces in 1973, the gun trucks continued to prowl the highways of Vietnam, protecting truckers and their cargo. With the end of the Vietnam War, the U.S. Army turned its attention from conducting major counterinsurgency operations. Instead, the

---

[25] HQ, 54th Transportation Bn (Truck), Operational Report of the 54th Transportation Battalion (Truck) WFR6AA, for the Period Ending 31 October 1968, 6SFOR-65 (R-1), 3 Nov 68.

[26] U.S. Congress, Senate, Committee on Veterans' Affairs, *Medal of Honor Recipients: 1863–1978* (Washington, D.C.: Government Printing Office, 1979).

Army concentrated on a possible ground war in Europe against Soviet armored columns, a battlefield with front lines, which obviated the need for gun trucks. As the years passed and the number of Vietnam veterans decreased in active service, the institutional memory of the Vietnam gun trucks faded. Almost forty years later, in Iraq, the U.S. Army truckers found themselves in a situation similar to that of their Vietnam predecessors. Once again, truckers converted cargo vehicles into armored fighting vehicles.

Today, "Eve of Destruction"—the only remaining example of the Vietnam gun trucks—rests quietly among the static displays at Fort Eustis, Virginia. Converted from a five-ton cargo truck of the 523d Transportation Company, Eve is armed with four .50-caliber machine guns. The truck provided daily route security in the Central Highlands and along the coast for three years before participating in the Cambodian incursion. During that operation, the vehicle escorted convoys from Qui Nhon to the Cambodian border, and the entire crew earned the Bronze Star for outstanding performance in protecting supply columns from enemy attacks. In January 1971, Eve led elements of the 8th Transportation Group north into I Corps Tactical Zone to participate in Operation Lam Son 719, the South Vietnamese invasion of Laos. Day and night, convoys to Khe Sanh and the Laotian border exposed Eve to numerous enemy attacks during the operation, but the gun truck never failed in its mission. Eve made its final run on 8 June 1971. Aware of the historical importance of the vehicle, the commander of the 523d Transportation Company had the vehicle shipped to the Transportation Museum at Fort Eustis. Now, far from the sounds of battle, the armored truck silently reminds those who view it of the innovation and old-fashioned American ingenuity of a previous generation of truckers, and the courage and sacrifice of transportation personnel who fought and sometimes died on the highways of Vietnam.

Some Observations on Americans Advising Indigenous Forces

Robert D. Ramsey III

In February 2006, the Combat Studies Institute (CSI) initiated a review of past United States Army efforts to build foreign armies. A month later, when the Department of Defense published its *Quadrennial Defense Review Report* and the Army began to centralize adviser training at Fort Riley, Kansas, CSI decided to refocus the study to field advisers in Korea, Vietnam, and El Salvador.[1] With the proliferation of advisory teams in Afghanistan and Iraq, the intent of the study was to investigate advisory duty from the perspective of advisers who worked face-to-face with host-nation personnel at the lowest levels on a daily basis—specifically, those serving in combat units or in pacification positions. CSI published this study as Occasional Paper (OP) 18, *Advising Indigenous Forces: American Advisors in Korea, Vietnam, and El Salvador,* in early September 2006.[2] In October, CSI followed with the publication of OP 19, *Advice for Advisors: Suggestions and Observations from Lawrence to the Present*, an anthology of readings from selected research materials gathered in the writing of *Advising Indigenous Forces*.[3] The following comments are based on these two research efforts.

To gain an understanding of the challenges that advisers face, *Advising Ingenious Forces* surveyed three American advisory experiences—Korea, Vietnam, and El Salvador. Korea, as a first major combat advisory effort, provided insights from a three-year conventional war experience working with the newly created South Korean Army. The United States Military Advisory Group to the Republic of Korea (KMAG) reached a maximum strength of 2,866 in 1953.[4] Vietnam, the largest and longest American advisory effort, provided a twelve-year combat and counterinsurgency experience working with a weak, but combat-experienced, South Vietnamese Army. Military Assistance

[1] Department of Defense, *Quadrennial Defense Review Report* (Washington, D.C.: Department of Defense, 6 February 2006).

[2] Robert D. Ramsey III, *Advising Indigenous Forces: American Advisors in Korea, Vietnam, and El Salvador* (Fort Leavenworth, Kans.: Combat Studies Institute Press, 2006).

[3] Robert D. Ramsey III, *Advice for Advisors: Suggestions and Observations from Lawrence to the Present* (Fort Leavenworth, Kans.: Combat Studies Institute Press, 2006).

[4] Alfred H. Hausrath, *The KMAG Advisor: Roles and Problems of the Military Advisor in Developing an Indigenous Army for Combat Operations in Korea* (Chevy Chase, Md.: Johns Hopkins University Operations Research Office, February 1957), p. 95.

Command, Vietnam (MACV), advisers served with combat units and on pacification advisory teams at the province and district level and on mobile advisory teams (MATs). To fill MACV adviser positions in 1968 required the Army to provide over seven division equivalents of officers and senior noncommissioned officers.[5] At a maximum strength of 14,332 advisers in 1970, MACV had almost 1,000 advisers working with regimental-and-below combat units and almost 8,000 advisers serving with pacification advisory teams.[6] El Salvador permitted a look at advising a limited, long-term counterinsurgency effort with an army without combat experience. Limited to fifty-five advisers in country, the six brigade-level operations, plans, and training teams (OPATTs) required a total of eighteen advisers.[7] Korea, Vietnam, and El Salvador each offered a unique experience; yet, the fundamental problem faced by field advisers—how to establish and maintain an effective working relationship with their counterparts to improve military effectiveness in addressing host-nation security problems—remained constant.

Adviser roles and duties evolved, particularly in Vietnam. For combat unit advisers—even KMAG—training, teaching, coaching, liaison, observing, tactical advising, and providing combat support were relatively straightforward, basic military tasks complicated by linguistic, cultural, and institutional challenges. After the increase of American military personnel in Vietnam in 1965, MACV combat unit advisers focused on coordinating American combat support assets, being liaisons with American units, and monitoring the status of their Vietnamese units. Over time, the tactical advisory effort in Vietnam evolved from training to tactical advice to combat support.[8] Counterinsurgency civil-military duties proved a complex, difficult task for pacification advisers in Vietnam—something often beyond their training, experience, and expertise. In 1990, the monitoring of human rights abuses had become an OPATT responsibility in El Salvador.[9] In each of the case studies, monitoring the status and actions of host-nation forces became an implied, if not specified, task.

At the lower levels, American military assistance and advisory groups (MAAGs) had to obtain results with a minimal number of personnel—normally five or less per team. Although regulations authorized advisers at the infantry battalion level, personnel shortages limited KMAG advisers to the infantry regiment and above.[10] The standard KMAG regimental advisory team consisted

[5] Bruce Palmer Jr., *The 25 Year War: America's Military Role in Vietnam* (Lexington: University of Kentucky Press, 1984), p. 178.

[6] Jeffrey J. Clarke, *Advice and Support: The Final Years, 1965–1973* (Washington, D.C.: U.S. Army Center of Military History, 1988), pp. 372–73.

[7] Cecil E. Bailey, "OPATT: The U.S. Army SF Advisers in El Salvador," *Special Warfare* (December 2004): 21.

[8] Clarke, *Advice and Support: The Final Years*, pp. 60–61.

[9] Mark A. Meoni, "The Advisor: From Vietnam to El Salvador" (Master's thesis, U.S. Army Command and General Staff College, 1992), p. 90.

[10] Robert K. Sawyer, *Military Advisors in Korea: KMAG in Peace and War* (Washington, D.C.: U.S. Army Center of Military History, 1962), p. 58.

of two officers and two noncommissioned officers.[11] In Vietnam, the infantry regiment advisory team consisted of one officer and two noncommissioned officers. Commanded by a captain, MACV infantry battalion advisory teams (and initially the district advisory teams) were authorized two officers and three noncommissioned officers.[12] Working to upgrade Vietnamese local security forces as part of the pacification effort, MATs commanded by a captain had five Americans and two Vietnamese, one as an interpreter.[13] In El Salvador, after mid-1985, the brigade OPATT consisted of a combat arms major and two senior Special Forces personnel—noncommissioned or warrant officers.[14] By design, commanders of advisory teams were at least one rank below the host-nation commander they advised. Because of personnel shortages, it was common for advisory team commanders to be a rank below that authorized for the position. As an example, *Once a Warrior King* is a firsthand account of an infantry first lieutenant in Vietnam who commanded a MAT, in lieu of a captain, before becoming a district senior adviser in lieu of a major.[15] Although most Americans were unconcerned about these differences in rank, the same was not true for their host-nation counterparts, who were extremely sensitive to rank differences.

While American advisory teams seem small given their tasks, a British counterinsurgency expert believed in the "need on the military side to keep the presence of foreign military advisors to the minimum. If things are not going right, it is most unlikely that the solution will be found merely by increasing the quantity of advisors. This is liable to be counterproductive and can reach the point at which advice begins to revolve on a closed circuit."[16] Often, more advice proved less effective, particularly when it was more of the same provided by inexperienced advisers who frequently rotated in and out of positions because of short tours of duty or assignment policies. Field advisory teams were small, frequently understrength, and normally filled with less than fully qualified members.

One would think that the limited number of advisers at each level and the numerous tasks expected of them during wartime—not to mention the alien environment in which they worked—would require some special selection process to identify those qualified for advisory duty. That was not the case. With the exceptions of province and district senior advisers for a short time toward the end of the Vietnam conflict, officers selected

[11] Hausrath, *The KMAG Advisor*, p. 96.

[12] Clarke, *Advice and Support: The Final Years*, p. 57; HQ, U.S. Military Assistance Command, Vietnam (MACV), Command History, 1965, 20 Apr 66, pp. 76–77. Hereafter cited as MACV, Command History, date.

[13] MACV, Command History, 1970, vol. 2, p. VII-67.

[14] Bailey, "OPATT: The U.S. Army SF Advisers in El Salvador," p. 21.

[15] David Donovan, *Once a Warrior King: Memories of an Officer in Vietnam* (New York: Ballantine, 1985).

[16] Robert Thompson, *Defeating Communist Insurgency: The Lessons of Malaya and Vietnam* (New York: Praeger, 1966), p. 165.

for the Military Assistance Officer Program (MAOP), and some advisers in El Salvador, the Army considered advisory duty a routine assignment that anyone could perform.[17] If a person met rank and branch qualification requirements and was eligible for an overseas tour, he was suited for advisory duty. If he volunteered, so much the better. In Korea, KMAG attracted few volunteers, but, in Vietnam prior to 1965, volunteers were common. With the buildup of American forces in 1965, increased demand for branch-qualified personnel and lower priority for advisory duty meant—just as it had in Korea—that inexperienced but enthusiastic junior officers often filled advisory positions. When demand exceeded supply, the Army waived the rank, branch, and experience requirements. In October 1970, two years after the establishment of captain-commanded MATs, only eighty captains headed the 487 teams.[18] In El Salvador, where adviser requirements emphasized Special Forces experience, prior service in the region, and Spanish-language capability, the general policy was that anyone who spoke Spanish and had served in Latin America could do advisory duty. Even when the military group in El Salvador (MILGRP) reduced the language requirement, the demand for five qualified OPATT chiefs exceeded the capacity of the Special Forces branch in 1990.[19] Advisory duty, despite comments from senior leaders, was never a top priority.

Only in Vietnam did advisers attend formal training courses prior to their assignments. In Korea, branch qualification was considered adequate. In El Salvador, language skills, regional experience, and a Special Forces background qualified a soldier for advisory duty. During Vietnam, a six-week military assistance training and advisory (MATA) course taught at Fort Bragg, North Carolina, focused on language, culture, and advisory skills rather than military skills.[20] The MATA course merely introduced and familiarized advisers with the challenges they would face. A MATA course instructor spoke of his student experience: "I left the MATA course knowing that this was going to be a very different experience . . . [and] prepared . . . not to expect everything I would face, but to expect that I was going to be immersed in a very different culture and adapting to that culture and understanding it was going to be complicated."[21] Provincial senior advisers, district senior

[17] Department of the Army (DA), AR 614–134, *Military Assistance Officer Program (MAOP)* 20 June 1971; Edwin E. Erickson and Herbert H. Vreeland, 3d, *Operational and Training Requirements of the Military Assistance Officer* (McLean, Va.: Human Services Research, May 1971).

[18] MACV, Command History, 1970, vol. 2, p. VII-67.

[19] Bailey, "OPATT: The U.S. Army SF Advisers in El Salvador," pp. 21–22.

[20] U.S. Army Special Warfare School, Program of Instruction for Military Assistance Training Advisory Course (MATA) (Fort Bragg, N.C.: U.S. Army Special Warfare School, April 1962).

[21] Interv, Shelby Sears with Anthony C. Zinni, 29 Jun 2004, p. 6, John A. Adams '71 Center for Military History and Strategic Analysis Cold War Oral History Project, Virginia Military Institute, available at www.vmi.edu/archives/Adams_Center/ZinniAC/ZinniAC_intro.asp.

advisers, and those selected for MAOP—personnel involved directly in pacification—received longer language and special training toward the end of Vietnam. However, just as in Korea, "in both the cases of Vietnam and El Salvador, military planners were not worried about attendance to advisor courses that dealt with culture, language, or irregular warfare. It was the [MOS] schools that took priority." [22] The Army considered military skills sufficient preparation for advisory duty.

Not only was adviser training generally nonexistent or limited, in-country orientation and training for advisory personnel, with the exception of most advisers in Vietnam, proved haphazard. In Vietnam, unlike the other two cases, adviser field manuals and in-country handbooks provided guidance to advisers.[23] During the Korean War, an adviser commonly reported directly to his KMAG team without any orientation. One KMAG adviser described his transition: "The officer I replaced met me at the rail-head (4 hours behind the division) turned his jeep over to me and gave me directions to the Division CP."[24] In mid-1953, KMAG issued for the first time the "Ten Commandments for KMAG Advisors," the duties and responsibilities of KMAG advisers.[25] Even in El Salvador, at least one OPATT chief read his orientation folders, went directly to his unit, and did not see the military group (USMILGP) commander for months.[26]

Interestingly, it is almost impossible to find a comment from an adviser in the three case studies where an adviser felt tactically, technically, or militarily unprepared for his duties—even those duties one or two levels above his rank and experience. It appeared that branch qualification, combined with American self-confidence, met the military requirements faced by most advisers. However, almost to a man, advisers mentioned the challenges posed by linguistic, cultural, and host-nation institutional barriers. It was in these areas—critical to an adviser's effectiveness—that most advisers felt inadequately prepared. Although it is true that "there has never been a training program of instruction [POI] to prepare military advisors for duty that all those with an interest might agree was comprehensive and complete," it seemed clear that topics that enhanced situational understanding were considered more useful by field advisers than those dealing with military skills.[27]

---

[22] MOS stands for military cccupational specialty. Meoni, "The Advisor: From Vietnam to El Salvador," p. 142.

[23] DA Field Manual 31–73, *Advisor Handbook for Counterinsurgency*, April 1965; DA Field Manual 31–73, *Advisor Handbook for Stability Operations*, October 1967; HQ, MACV, *Handbook for Military Support of Pacification*, February 1968; HQ, MACV, *rf / pf Advisor Handbook*, January 1971.

[24] Hausrath, *The KMAG Advisor*, p. 41.

[25] Ibid., p. 15.

[26] David L. Shelton, "Some Advice for the Prospective Advisor," *Marine Corps Gazette* (October 1991): 57.

[27] Meoni, "The Advisor: From Vietnam to El Salvador," p. 184.

T. E. Lawrence wrote in *Seven Pillars of Wisdom*, "I was sent to these Arabs as a stranger, unable to think their thoughts or subscribe their beliefs, but charged by duty to lead them forward and to develop to the highest any movement of theirs profitable to England in her war."[28] So it was with American advisers who worked in an alien, often hostile, environment in pursuit of American objectives. Generally, they lacked language skills, cultural knowledge, understanding of the host nation, and knowledge of the host-nation military. Each of these shortcomings created obstacles to understanding. Yet, the service expected advisers to produce results. They were to develop rapport—a close personal relationship—and to provide useful advice to their counterparts. Few advisers understood that empathy (being able to understand a problem from the perspective of their counterparts) would be critical to becoming an effective adviser.[29] Fewer advisers possessed the knowledge, skills, and abilities to seek that perspective. A review of comments from advisers, their counterparts, and special studies provided insights into many of the challenges advisers confronted.

Advisers worked in an alien environment; one that shared few things in common with many Americans. Most advisers lacked language skills; as a result, they were deaf—they could not understand what was said around them. In Korea, no field KAMG adviser spoke Hangul fluently. Advisers depended on Korean translators. In Vietnam, most MACV advisers received some basic language training, but they remained tied to Vietnamese translators. Even in El Salvador, where Spanish was required, few advisers were native speakers. Lack of language skills impaired communication. Interpreters helped, but marred communication meant frequent misunderstandings that were difficult to sort out without linguistic ability. Observations that "to surmount the language barrier the American advisor had to be an inventive teacher, combining enthusiasm and knowledge with patience and tact," missed the point.[30] The language barrier, even using interpreters, remained a significant obstacle to effective advising.

Not only did the language barrier hinder advisers, the cultural barrier did too. What passed for cultural awareness was generally descriptive, superficial, and stereotypical. Inability to understand what they observed meant that advisers were partially blind. A MACV adviser observed, "We did not understand what was going on in Vietnam. We were in a foreign land among people of a different culture and mindset. . . . The information sent across the cultural divide was not the information received. There was a disconnect. One thing was said and another thing was heard. One thing was meant and another

[28] T. E. Lawrence, *Seven Pillars of Wisdom: A Triumph* (New York: Anchor Books, 1991), p. 30.

[29] Irwing C. Hudlin, "Advising the Advisor," *Military Review* (November 1965): 94–96.

[30] Historical Rpt, Walter G. Hermes, Survey of the Development of the Role of the U.S. Military Advisor, Office of the Chief of Military History, 1965, p. 82, U.S. Army Center of Military History files, Washington, D.C.

thing was understood. . . . Meaning, intent, and truth were lost in translation."[31] Even though "frequently the success of the advisor depended as much upon his behavior as upon his professional ability," most advisers remained unaware of the impact that their actions and inactions had on their counterparts.[32] At best, instruction and orientations made advisers aware of basic cultural issues. However, understanding the cultural *what* without knowing the more important cultural *how* and *why* locals did what they did created only an impression of understanding. Even with the training courses during Vietnam, an adviser found that the "linguistic and cultural barrier . . . was almost impossible for the advisor to breach"[33] Inadequate language skills and lack of cultural knowledge made it difficult for an adviser to comprehend accurately the local situation or to understand that situation from the perspective of his counterpart.

Advisers worked where American goals, capabilities, limitations, techniques, procedures, and doctrines—not to mention linguistic, cultural, and institutional imperatives—were not those of the host nation. Sir Robert Thompson, Malaya veteran and British adviser in Vietnam, noted, "It is essential, therefore, for the advisor to look at everything from the local point of view and not to expect that the provision of aid will do more than provide the very limited benefits for which it was intended. He cannot expect that the threatened country will either organize itself or conduct its affairs on the same lines, or in accordance with the same standards, as those of the supporting power. The real point here, which is all the advisor can hope for, is to get the local government to function effectively and at least to take the necessary action itself, even if it is done in its own traditional way."[34] Although this appears obvious, the fact remained that many field advisers and their leaders failed to understand or accept that the American way was not the best or the only way to do things.

Just as the host nation was not the United States, so the host-nation military was not the American military. Its roles, goals, procedures, capabilities, and limitations were different. Advisers needed to understand these differences and their implications for working with their counterparts. Although desirous of American training and resources, no host-nation military wanted to become a small clone of the American military. At best, it wanted to improve its combat effectiveness. In 1965, a RAND study recommended that "the advisor must learn to recognize and evaluate the relative role of his counterpart in the Vietnamese military's social structure, his freedom of action and of expression,

[31] Martin J. Dockery, *Lost in Translation: VIETNAM, A Combat Advisor's Story* (New York: Ballantine, 2003), p. 93.

[32] Hermes, Survey of the Development of the Role of the U.S. Military Advisor, pp. 82–83.

[33] Stuart A. Herrington, *Silence Was a Weapon: The Vietnam War in the Villages* (Novato, Calif.: Presidio Press, 1982), p. 191. (Available in paperback as *Stalking the Vietcong: Inside Operation Phoenix, A Personal Account*).

[34] Thompson, *Defeating Communist Insurgency*, p. 161.

before he can develop a useful working relationship with him."[35] Cultural and institutional differences were barriers that a senior MACV adviser suggested that advisers "must take . . . always into account. This seems like gratuitous advice . . . [but it was often] ignored by the complacent advisor who has his counterpart 'all figured out'—by American standards—and then is astounded when his counterpart does something 'entirely out of character.'"[36] It proved relatively simple to train tactical skills, but almost impossible to make deeper institutional changes without host-nation support, as provided in Korea. But even there in the end, just as in Vietnam and El Salvador, the Korean army remained uniquely Korean, despite its American structure, training, and equipment.

Because of the alien environment described above and the frequent turnover of advisers, many counterparts were resistant—indifferent at best, if not hostile—to their advisers' attempts to establish a close relationship. Without personal and institutional reasons to bond, and given the short-term focus of inexperienced junior American advisers serving brief assignments, rapport was easier to discuss than establish. Most host-nation commanders had combat experience, had worked with numerous advisers, had seniority in rank, and had command responsibility for their units. This meant that they seldom shared the enthusiasm, immediate focus, and myriad of projects, many of questionable utility, of their adviser. Many considered advisers arrogant and pushy. Host-nation commanders believed that mutual trust and respect, as well as an open, honest willingness to work together, were more important traits for an adviser than military competence.[37] Many did not find these qualities in their American advisers. To be effective, advisers needed interpersonal and intercultural skills more than military skills. When lower-level advisers provided their counterparts with something they needed—such as combat support assets in Korea and Vietnam, or pacification assets in Vietnam—a counterpart had an incentive to work with his adviser. Rapport, as noted by both advisers and counterparts, was not permanent. It could easily be damaged by a misunderstanding compounded by linguistic, cultural, personal, or institutional differences, as well as honest disagreement on an issue.

Rapport was merely a prerequisite for successful advising. It was, as an adviser noted, "only the prelude to his major objective: inspiring his counterpart to effective action."[38] This required the adviser to provide useful advice, something that made sense to his counterpart. For the host-nation commander, the advice had to fit his needs, be within his capabilities, be relevant to his specific problem, and be communicated

---

[35] Gerald C. Hickey, *The American Military Advisor and His Foreign Counterpart: The Case of Vietnam* (Santa Monica, Calif.: RAND Corp., March 1965), p. 14.

[36] Bryce F. Denno, "Advisor and Counterpart," *Army* (July 1965): 27–28.

[37] Hausrath, *The KMAG Advisor*, pp. 28–30.

[38] Denno, "Advisor and Counterpart," p. 26.

clearly and convincingly. When the counterpart understood, accepted, and was willing to act on the advice in a timely manner, the adviser had performed his job. For many advisers, this proved problematic, given inadequate understanding of the language, cultural issues, host-nation military institutional norms or procedures, and local conditions. Advisers often misunderstood or were ignorant of important things.

So what can we say about the American advisory efforts in Korea, Vietnam, and El Salvador? First, in each case, the response was ad hoc. When an unanticipated requirement arose, the Army established procedures, applied resources, obtained mixed results, and then forgot the experience. Second, advisory duty was a supporting effort that produced mixed results. Third, field advisory teams were small and frequently undermanned. Fourth, any soldier with the required military specialty and rank was considered qualified for advisory duty, although rank and other requirements were often waived. Fifth, language and culture training was not deemed critical for advisory duty—military skills were of primary importance. And last, advisory duty was a short-term, one-time assignment for most personnel. Despite these experiences and numerous special studies, each time a requirement for advisers arose, it was addressed as if it were a new phenomenon. This meant, in the words of a senior MACV adviser, that in each case, "All too often insight is gained too late, and through adverse experience."[39]

What can be taken away from a survey about these three American advisory efforts? First, it is a difficult job. Working effectively with indigenous forces may be the most difficult military task. Second, not everyone can do advisory duty well. Advisers should be carefully screened and selected. Third, advisers should receive in-depth training about the host nation, its history, its culture, and its language. Without such training, situational awareness is almost impossible. Fourth, adviser training should focus on advisory duties and harnessing host-nation institutions, organizations, procedures, capabilities, and limitations. It should not focus on U.S. solutions to host-nation problems. Fifth, longer, repetitive tours by specially selected and well-trained advisers enhance the development of rapport with host-nation counterparts. Lastly, no matter how capable field advisers are, their success depends upon the support structure established between the host nation and their U.S. chain of command.[40] Paraphrasing Lawrence's 1933 comment to Liddell Hart that "with 2,000 years of examples behind us we have no excuse, when fighting, for not fighting well,"[41] *Advising Indigenous Forces* concludes, "since the beginning of World War II, with over 65 years of American examples behind

---

[39] HQ, Delta Regional Assistance Command, Vietnam, Senior Officer Debriefing of Maj Gen John H. Cushman, RCS CSFOR-74 (14 January 1972), p. 2.

[40] Ramsey, *Advising Indigenous Forces*, pp. 114–17.

[41] Jeremy Wilson, *Lawrence of Arabia: The Authorized Biography of T. E. Lawrence* (New York: Atheneum, 1990), p. 908.

us, we have no excuse, when advising, for not advising well."[42] It remains to be seen if the U.S. military will continue to follow an ad hoc approach as it did in these three case studies, or if it will accept the lessons from past advisory efforts in order to be able to advise well.

[42] Ramsey, *Advising Indigenous Forces*, p. 118.

Contributors

Ted Ballard served as a historian with the U.S. Army Center of Military History from 1980 until his retirement in 2004. In 1984–1986, he was a member of the 116th Military History Detachment, Virginia Army National Guard, and in 1986–1989 he was the assistant division historian with the Virginia Defense Force. Ballard served in the U.S. Army from 1965 to 1967, spending 1967 in Vietnam with the 523d Transportation Company.

Frederic L. Borch retired from the Army after twenty-five years as a judge advocate and is now the regimental historian and archivist for the Army Judge Advocate General's Corps. He received his M.A. in history from the University of Virginia and his J.D. from the University of North Carolina.

Lt. Col. John A. Boyd, Ph.D., is the command historian for the 81st Regional Readiness Command, U.S. Army Reserve. He did his undergraduate work at Vanderbilt University, earned his M.Ed. and M.A. in history from the University of Cincinnati, and received his Ph.D. from the University of Kentucky in 1999. At the University of Kentucky, he specialized in Civil War–era sectionalism and did minor work in Middle East history. Boyd's first military deployment as a historian was in support of Operation JOINT ENDEAVOR, the U.S.-NATO effort in Bosnia. His most recent deployment (2005–2006) was as the command historian for the Multi-National Force-Iraq, Baghdad.

Richard Carrier attended Laval University where he obtained his Ph.D. in international history. From 1996 to 2007, he was a senior lecturer at the Saint-Jean Campus of the Royal Military College of Canada. In January 2008, he became assistant professor at the Royal Military College of Canada in Kingston, where he teaches courses on the history of the Second World War and on the history of international relations between 1945 and 1991. Carrier's research interests mainly focus on the history and sociology of military institutions, the history of war in Europe since 1789, and the political and military history of Italy in the twentieth century.

Kendall D. Gott retired from the U.S. Army in 2000 having served as an armor and military intelligence officer. His combat experience consists of the Persian Gulf War and two bombing campaigns against Iraq. A native of Peoria, Illinois, he received his B.A. in history from Western Illinois University, and a master's of military art and science from the U.S. Army Command and General Staff College. Before returning to Kansas in 2002, he was an adjunct professor of history at Augusta State University and the Georgia Military College. Gott joined the staff of the U.S. Army Combat Studies Institute at Fort Leavenworth, Kansas, in October 2002. As a senior historian, he heads the Research and Publications Team.

Chris Madsen is an associate professor in the Department of Defence Studies at the Royal Military College of Canada and the Canadian Forces College in Toronto. He obtained a Ph.D. in history from the University of Victoria in 1996 and was a Department of National Defence and Social Sciences and Humanities Research Council of Canada postdoctoral fellow at the University of Calgary until 1999. He currently teaches strategy and history to professional military officers in a staff college environment. Madsen has authored and edited several books and journal articles, including *Another Kind of Justice: Canadian Military Law from Confederation to Somalia* (University of British Columbia Press, 1999) and *Kurt Meyer on Trial: A Documentary Record* (Canadian Defence Academy Press, 2007). His latest work is titled *Military Law and Operations*.

Robert M. Mages is the chief of Oral History, U.S. Army Military History Institute, Army Heritage and Education Center, Carlisle Barracks, Pennsylvania. He holds an M.A. in modern European history from Washington State University.

Maj. (Ret.) Steven J. Rauch has served as the U.S. Army Signal Center command historian since August 2002. He received his M.A. in history from Eastern Michigan University and holds an M.S. in adult and continuing education from Kansas State University. He specializes in American colonial and early national U.S. military history, nineteenth-century European history, and general military history. Rauch served as an assistant professor of history at the U.S. Army Command and General Staff College, where he taught the history of modern warfare and early U.S. military history. He served in the U.S. Army as an ordnance officer and retired from active duty in 2002.

Robert D. Ramsey III, a historian with the Combat Studies Institute at Fort Leavenworth, retired from the U.S. Army in 1993 after twenty-four years of service as an infantry officer that included tours in Vietnam, Korea, and the Sinai. He earned an M.A. in history from Rice University. Ramsey taught military history for three years at the United States Military Academy and six years at the U.S. Army Command and General Staff College. His publications include four Combat Studies Institute occasional papers

(OP)—OP 18, *Advising Indigenous Forces: American Advisors in Korea, Vietnam, and El Salvador*; OP 19, *Advice for Advisors: Suggestions and Observations from Lawrence to the Present*; OP 24, *Savage Wars of Peace: Case Studies of Pacification in the Philippines, 1900–1902*; and OP 25, *Masterpiece of Counterguerrilla Warfare: BG J. Franklin Bell in the Philippines, 1901–1902.*

Lt. Col. (Ret.) Mark J. Reardon, a senior historian at the U.S. Army Center of Military History, is the author of *Victory at Mortain: Stopping Hitler's Panzer Counteroffensive* (University Press of Kansas, 2003), and coauthor of *American Iliad: The 18th Infantry Regiment in World War II* (Aberjona Press, 2004) and *From Transformation to Combat: The First Stryker Brigade at War* (U.S. Army Center of Military History, 2007)), as well as numerous magazine articles on subjects ranging from Lewis and Clark's expedition to World War II in Europe. Reardon has appeared as a guest commentator on Oliver North's "War Stories," the History Channel, and the Military Channel.

Stephen I. Rockenbach is an assistant professor at Virginia State University in Petersburg, where he teaches courses on American history. He received his M.A. in history in 2000, and his Ph.D. in history in 2005, both from the University of Cincinnati. His publications include "A Border City at War: Louisville and the 1862 Confederate Invasion of Kentucky," in *Ohio Valley History*, several entries in *The Encyclopedia of Northern Kentucky*, seven essays in the *Gale Library of Daily Life: American Civil War* (Gale Cengage, 2008), and numerous book reviews. His research interests include the border region during the Civil War, guerrilla warfare, emancipation, and the memory of the Civil War. Rockenbach is currently revising a manuscript that compares the wartime experience of citizens in Corydon, Indiana, and Frankfort, Kentucky.

Russell G. Rodgers has served with armor, infantry, and military intelligence during a varied U.S. Army career. He received his master's degree with honors in land warfare from the American Military University and is currently an Army staff historian and adjunct professor at American Military University. He is the author of *Historic Photos of General George Patton* (Turner Publishing, 2007) and has contributed a chapter on suicide bombers in *Terrorism in the 21st Century: Unanswered Questions* (Praeger, Fall 2008). Rodgers is currently finishing a book on the asymmetric insurgency doctrine of the Prophet Muhammad.

Eric A. Sibul is in charge of military theory and military history programs at the Baltic Defence College in Tartu, Estonia. He is currently working on his Ph.D. dissertation at the University of York in the United Kingdom. His dissertation examines rail transport in the Korean War and the relationship between the U.S. Army Transportation Corps and the Korean National Railroad. Sibul previously taught history and business administration at Central Texas College and Kutztown University of Pennsylvania.

Gregory J. W. Urwin holds his Ph.D. from the University of Notre Dame. He is a professor of history and associate director of the Center for the Study of Force and Diplomacy at Temple University. He has authored or edited eight books, including *Custer Victorious: The Civil War Battles of General George Armstrong Custer* (University of Nebraska Press, 1983), *Black Flag over Dixie: Racial Atrocities and Reprisals in the Civil War* (Southern Illinois University Press, 2004), pictorial histories of the U.S. Cavalry and Infantry, and the prize-winning *Facing Fearful Odds: The Siege of Wake Island* (University of Nebraska Press, 2002). Urwin's ninth book will explain why the Wake Island defenders emerged from prison camp with such a high survival rate. After that, he plans to write a social history of Cornwallis' 1781 Virginia campaign.

Thijs W. Brocades Zaalberg is a research associate at the Netherlands Institute for Military History in The Hague. He previously worked as a defense analyst at the Clingendael Centre for Strategic Studies and is the author of *Soldiers and Civil Power: Supporting or Substituting Civil Authorities in Modern Peace Operations* (Amsterdam University Press, 2006). He coauthored *American Visions of the Netherlands-East Indies: U.S. Foreign Policy and Indonesian Nationalism, 1920–1949* (Amsterdam University Press, 2002). Brocades Zaalberg earned his Ph.D. in history from the University of Amsterdam.